MICROBIOLOGY RESEARCH ADVANCES

BETA-LACTAMASES

AN OVERVIEW

MICROBIOLOGY RESEARCH ADVANCES

Additional books and e-books in this series can be found on Nova's website under the Series tab.

MICROBIOLOGY RESEARCH ADVANCES

BETA-LACTAMASES

AN OVERVIEW

THOMAS E. STEAD
EDITOR

Copyright © 2020 by Nova Science Publishers, Inc.

All rights reserved. No part of this book may be reproduced, stored in a retrieval system or transmitted in any form or by any means: electronic, electrostatic, magnetic, tape, mechanical photocopying, recording or otherwise without the written permission of the Publisher.

We have partnered with Copyright Clearance Center to make it easy for you to obtain permissions to reuse content from this publication. Simply navigate to this publication's page on Nova's website and locate the "Get Permission" button below the title description. This button is linked directly to the title's permission page on copyright.com. Alternatively, you can visit copyright.com and search by title, ISBN, or ISSN.

For further questions about using the service on copyright.com, please contact:
Copyright Clearance Center
Phone: +1-(978) 750-8400   Fax: +1-(978) 750-4470   E-mail: info@copyright.com.

NOTICE TO THE READER

The Publisher has taken reasonable care in the preparation of this book, but makes no expressed or implied warranty of any kind and assumes no responsibility for any errors or omissions. No liability is assumed for incidental or consequential damages in connection with or arising out of information contained in this book. The Publisher shall not be liable for any special, consequential, or exemplary damages resulting, in whole or in part, from the readers' use of, or reliance upon, this material. Any parts of this book based on government reports are so indicated and copyright is claimed for those parts to the extent applicable to compilations of such works.

Independent verification should be sought for any data, advice or recommendations contained in this book. In addition, no responsibility is assumed by the Publisher for any injury and/or damage to persons or property arising from any methods, products, instructions, ideas or otherwise contained in this publication.

This publication is designed to provide accurate and authoritative information with regard to the subject matter covered herein. It is sold with the clear understanding that the Publisher is not engaged in rendering legal or any other professional services. If legal or any other expert assistance is required, the services of a competent person should be sought. FROM A DECLARATION OF PARTICIPANTS JOINTLY ADOPTED BY A COMMITTEE OF THE AMERICAN BAR ASSOCIATION AND A COMMITTEE OF PUBLISHERS.

Additional color graphics may be available in the e-book version of this book.

Library of Congress Cataloging-in-Publication Data

Names: Stead, Thomas E., editor.
Title: Beta-lactamases : an overview / [edited by] Thomas E. Stead.
Description: New York : Nova Science Publishers, [2019] | Series:
  Microbiology research advances | Includes bibliographical references and
  index. |
Identifiers: LCCN 2019053629 (print) | LCCN 2019053630 (ebook) | ISBN
  9781536168181 (paperback) | ISBN 9781536168198 (adobe pdf)
Subjects: LCSH: Beta lactamases.
Classification: LCC QP609.B46 B474 2019 (print) | LCC QP609.B46 (ebook) |
  DDC 579.3/5--dc23
LC record available at https://lccn.loc.gov/2019053629
LC ebook record available at https://lccn.loc.gov/2019053630

Published by Nova Science Publishers, Inc. † New York

Contents

Preface		vii
Chapter 1	The Emergence and Evolution of Extended-Spectrum β-Lactamases (ESBL) and AmpC in Food-Producing Animals, Farm Environment and Their Products *Jungwhan Chon, Dongryeoul Bae, Saeed Khan and Kidon Sung*	1
Chapter 2	Resistance Mechanisms in *Acinetobacter* with an Emphasis on β-Lactamases: An Update *Aparna Shivaprasad and Beena Antony*	119
Chapter 3	Golden Age of β-Lactam Antibiotics to Antibiotic Resistance *Ritika Chauhan and Jayanthi Abraham*	143
Index		169
Related Nova Publications		179

PREFACE

In this compilation, the authors describe the occurrence and characteristics of ESBL/AmpC-producing Enterobacteriaceae in poultry, cattle and pigs, pointing to risk factors that lead to their spread and highlighting possible mitigation strategies that could be applied to reduce their prevalence in food-producing animals.

One review focuses on the wide array of antimicrobial resistance mechanisms that have been described in A. baumannii. The most concerning ones being the β-lactamases with carbapenemase activity, i.e, Ambler class-D serine oxacillinases and Ambler class-B metallo-β-lactamases.

The closing study will walk readers through the discovery and golden age of β-lactam, its mechanism of action as broad-spectrum antibiotic, the expansion of beta-lactam derivatives and β-lactamase inhibitors to antibiotic resistance in β-lactams.

Chapter 1 - β-Lactam antimicrobials have been used for the treatment of bacterial infections in both human and food producing animals. Enterobacteriaceae resistant to the β-Lactam antimicrobials have been recovered increasingly from food producing animals and meat products and might play a role as potential source for spreading antibiotic resistant genes to the surrounding environment. A variety of β-lactamases, such as ESBL (CTX-M, TEM, SHV) and AmpC (CMY, DHA-1, ACT-1), have been identified in *Escherichia coli* and non-typhoidal *Salmonella* from food-producing animals. The use of cephalosporins and other antimicrobials in animal farms is believed to contribute to the enrichment and dissemination of *E. coli* or *Salmonella* strains encoding ESBL/AmpC genes. The detection of *E. coli* O25-ST131 strains, a human pathogen and encoding CTX-M-15 in sick animals, has caused serious concerns about its spread and zoonotic transfer. Presence of mobile genetic elements, integrons, insertion sequences, transposons and plasmids in ESBL/AmpC producers aids in the transfer of these genes and seriously limits the options for antimicrobial therapy. This review describes the occurrence and characteristics of ESBL/AmpC-producing Enterobacteriaceae in poultry, cattle and pigs, points to risk factors that lead to their spread and highlights possible mitigation strategies that could be applied to reduce their prevalence in food producing animals.

Chapter 2 - *A. baumannii* has emerged as a ubiquitous 'Superbug' causing both community and health care associated infections in the recent past. More than 40 genospecies of Acinetobacter have been identified using Multilocus sequence analysis. Among them *A. baumannii, A. pittii, A. nosocomialis, A. seifertii* and *A. dijkshoorniae* that forms the AB complex are more clinically relevant. A wide array of antimicrobial resistance mechanisms have been described in *A. baumannii*. The most concerning ones being the β-lactamases with carbapenemase activity, i.e., Ambler class-D serine oxacillinases (OXA-type) and Ambler class-B metallo-β-lactamases (MBLs). More recently, *A.baumannii* is also known to possess New Delhi Metallo β-lactamase (NDM), a broad-spectrum MBL, renamed as Plasmid encoded carbapenemase resistant metallo β-lactamase (PCM).

In our personal experience of analysing 888 strains of *Acinetobacter* isolated from various clinical samples over a period of 5 years, 80.5%, were extremely drug resistant *A. baumannii* (XDR-AB), showing resistance to multiple classes of antibiotics including carbapenems and 75% of them produced biofilm. Molecular characterization of 251 isolates done by PCR revealed genes such as bla_{OXA-51} (90.4%), bla_{OXA-58} (58.2%), bla_{OXA-23} (57.4%). 41% of our isolates were NDM-1 producers. Co-existence of multiple MBL and carbapenemase genes in clinical isolates of *A. baumannii* is a potential threat in hospitals and is a problem to reckon with. It should be seriously considered and addressed with alternative and newer therapeutic strategies, strict infection control measures and continuous surveillance. Initial screening of the putative carbapenemase producers will help to organize intervention and early directed therapy.

Chapter 3 - Beta-lactam antibiotics are a group of antibiotics that hold β-lactam ring in molecular structure. These are widely used antibiotics mainly in developing countries against bacterial infections. β-*lactam* group of antibiotics, originally found in fungi *Penicillium notatum* inhibit cell wall biosynthesis in bacteria especially Gram-positive organisms. The development of new derivatives of β-lactam class- carbapenem, cephalosporins, cephamycin, monobactam are efficient against bacterial species which have developed resistance mechanisms. This review will walk readers through the discovery and golden age of β-lactam, mechanism of action as broad-spectrum antibiotic, expansion of beta-lactam derivatives, β-lactamase inhibitors to antibiotic resistance in β-lactams.

In: Beta-Lactamases: An Overview
Editor: Thomas E. Stead

ISBN: 978-1-53616-818-1
© 2020 Nova Science Publishers, Inc.

Chapter 1

THE EMERGENCE AND EVOLUTION OF EXTENDED-SPECTRUM β-LACTAMASES (ESBL) AND AmpC IN FOOD-PRODUCING ANIMALS, FARM ENVIRONMENT AND THEIR PRODUCTS

Jungwhan Chon[1], PhD, Dongryeoul Bae[2], PhD, Saeed Khan[1], PhD and Kidon Sung[1],*, PhD

[1]Division of Microbiology, National Center for Toxicological Research, Jefferson, AR, US
[2]Office of Regulatory Affairs, Arkansas Regional Laboratory, US Food and Drug Administration, Jefferson, AR, US

---

* Corresponding Author's Email: Kidon.Sung@fda.hhs.gov.

ABSTRACT

β-Lactam antimicrobials have been used for the treatment of bacterial infections in both human and food producing animals. Enterobacteriaceae resistant to the β-Lactam antimicrobials have been recovered increasingly from food producing animals and meat products and might play a role as potential source for spreading antibiotic resistant genes to the surrounding environment. A variety of β-lactamases, such as ESBL (CTX-M, TEM, SHV) and AmpC (CMY, DHA-1, ACT-1), have been identified in *Escherichia coli* and non-typhoidal *Salmonella* from food-producing animals. The use of cephalosporins and other antimicrobials in animal farms is believed to contribute to the enrichment and dissemination of *E. coli* or *Salmonella* strains encoding ESBL/AmpC genes. The detection of *E. coli* O25-ST131 strains, a human pathogen and encoding CTX-M-15 in sick animals, has caused serious concerns about its spread and zoonotic transfer. Presence of mobile genetic elements, integrons, insertion sequences, transposons and plasmids in ESBL/AmpC producers aids in the transfer of these genes and seriously limits the options for antimicrobial therapy. This review describes the occurrence and characteristics of ESBL/AmpC-producing Enterobacteriaceae in poultry, cattle and pigs, points to risk factors that lead to their spread and highlights possible mitigation strategies that could be applied to reduce their prevalence in food producing animals.

Keywords: extended-spectrum β-lactamases (ESBL), AmpC, β-lactamase, food-producing animals, Enterobacteriaceae

INTRODUCTION

Last decade has seen a tremendous increase in the presence of ß-lactamases in gram-negative bacteria. These can be broadly classified into two groups: extended-spectrum ß-lactamases (ESBLs) and AmpC ß-lactamases (AmpCs) conferring resistance to a wide range of ß-lactam antibiotics (Thomson, 2010). ESBL class of lactamases (TEM, SHV, and CTX-M class) confer resistance to the first-, second-, third-, and fourth generation cephalosporins, monobactams, and penicillins but not to the carbapenems or the cephamycins (Gutkind et al., 2013). In the late 1990s, the most common type of ESBLs in *Klebsiella* species and *E. coli* in hospital and community settings were SHV and TEM types (Chong et al., 2011). After the introduction of third and fourth generation cephalosporins, the epidemiology of ESBL/AmpC carriers and ESBL/AmpC genes has changed dramatically (Pitout, 2013). The presence of most ESBL class of lactamases on mobile genetic elements, transposons, and conjugative plasmids makes their transmission more efficient (Canton et al., 2008). They are now found worldwide in food-producing animals (poultry, cattle, pigs) and pose tremendous risk to public health (Madec et al., 2017; Rubin and Pitout, 2014). AmpC β-lactamases, on the other hand, are present on the chromosome and make the bacteria resistant to first-, second-, and third-generation cephalosporins, penicillins and cephamycins, but not to fourth-generation cephalosporins and carbapenems (Jacoby, 2009). The AmpC β-lactamases, mostly CMY and its variants, have now moved onto plasmids as well (Hawkey and Jones, 2009). Since most of the transposable elements and conjugative plasmids carry antimicrobial resistance genes to multiple antibiotics, combination of ESBLs/AmpC and multidrug resistance makes them a grave threat and risk to public health.

ESBL/AmpC Producers in Food Producing Animals

Poultry

E. coli and *Salmonella* account for most of the ESBL and AmpC-producing bacteria in poultry (Saliu et al., 2017). From healthy chickens, ESBL-producing *E. coli* and AmpC-producing *S. enterica* isolates were first reported in Spain in 2003 and Canada in 2002, respectively (Table 1) (Allen and Poppe, 2002; Brinas et al., 2003). The prevalence of ESBL/AmpC-encoding *E. coli* recovered from broiler chickens in Czech Republic, Portugal, and Germany was 1.9%, 42.1%, and 67.1%, respectively (Costa et al., 2009; Kolar et al., 2010; Laube et al., 2013). Comparative studies on the recovery rate of ESBL-producing bacteria between broiler and layer chickens have been done in the Netherlands and India, and the results were consistent. Broiler chickens (81-87.4%) carried more bacteria harboring ESBL than layer (42.1-65.0%) in both countries (Blaak et al., 2015; Brower et al., 2017). From sick poultry in Germany, a low occurrence (0.8%, 20/2,391) of ESBL-producing *E. coli* were reported between 2008 and 2014 (Michael et al., 2017). A comparative investigation of ESBL-containing gram-negative bacteria between conventional and organic chicken meat in the Netherlands indicated that almost 100% of the conventional chicken meat samples possessed ESBL, and the organic chicken meat samples also exhibited a high prevalence of ESBL producers (84.0%) (Cohen Stuart et al., 2012). The distribution of major ESBL genes and resistance to other antibiotics was similar between Netherland and German conventional and organic meat specimens (Kola et al., 2012). From healthy and sick chicken, and poultry products, high detection rate of ESBL/AmpC-expressing Enterobacteriaceae was observed in Germany (67.1%), China (60.8%), and the Netherlands (94.0%), respectively, while low detection rate was detected in Canada (0.1%), Egypt (6.0%), and Switzerland (0.2%), respectively (Allen and Poppe, 2002; El-Shazly et al., 2017; Hindermann et al., 2017; Laube et al., 2013; Voets et al., 2013; Yuan et al., 2009). More *E. coli* isolates encoding β-lactamase genes were found from sick chickens (7.0% in India and 28.7% in Nigeria) than healthy chickens (2.6% in India and 4.3% in Nigeria) (Kar et al., 2015; Mamza et al., 2010). Among healthy chicken and meat samples, 0-60.0% of healthy chicken and 1.4-28.9% of the meat specimens from Columbia, Ghana, Japan, Nigeria and Portugal contained ESBL/AmpC-carrying Enterobacteriaceae (Castellanos et al., 2017; Hiroi et al., 2011; Machado et al., 2008; Ojo et al., 2016; Rasmussen et al., 2015).

Table 1. Presence of ESBL/AmpC-producing Enterobacteriaceae in poultry

Species	Sample source	Year	Country	% (Sample #)	CTX-M	TEM	SHV	AmpC and other β-lactamase	Other resistance Phenotype	Other resistance Genotype	MLST	Inc group	Reference
S. Keurmassar	Meat	2000	Senegal	5 Salmonella strains			12						(Cardinale et al., 2001)
S. Bredeney, S. Ohio	Healthy	1994-1999	Canada	0.1 (8426)		1		CMY-2	CHL, FLO, GEN, KAN, NEO, SPE, STR, SUL, TET, TOB	aadB			(Allen and Poppe, 2002)
E. coli	Healthy	2000-2001	Spain	4.2 (120)	14	1b	12	CMY-2, AmpC	CHL, CIP, GEN, KAN, NAL, STR, SXT, TET, TOB				(Brinas et al., 2003)
S. Virchow	Healthy/meat	2002-2003	France	9 ESBL-producing Salmonella spp.	9	TEM			KAN, NAL, SPE, STR, SUL, TET, TMP				(Weill et al., 2004)
E. coli	Meat	2002	Taiwan	14.0 (100 meat samples)				CMY-2	CHL, CIP, GEN, LEV, NAL, SXT, TOB				(Yan et al., 2004)
E. coli	Healthy	2002	Taiwan	3.1 (32 fecal samples)				CMY-2	CHL, SXT				(Yan et al., 2004)
S. Bareilly, S. Blockley, S. Braenderup, S. Infantis, S. Paratyphi B, S. Typhimurium, S. Virchow	Healthy/meat	2001-2002	Netherlands	13 ESBL-/AmpC-producing Salmonella	2	1, 20, 52	2	ACC-1	GEN, NAL, NEO, SPE, STR, SUL, TET, TMP				(Hasman et al., 2005)

Table 1. (Continued)

Species	Sample source	Year	Country	% (Sample #)	CTX-M	TEM	SHV	AmpC and other β-lactamase	Other resistance	MLST	Inc group	Reference
S. Blockley, S. Paratyphi B var. Java, S. Typhimurium	Meat	2001-2002	Netherlands	7 ESBL-producing Salmonella spp.		20, 52	12		SPE, SUL, TMP			(Hasman et al., 2005)
E. coli	Healthy	1999-2002	Japan	1.7 (1082)	2, 18	1		CMY-2, AmpC	CHL, ENR, KAN, NAL			(Kojima et al., 2005)
S. Virchow	Meat	2001	Greece	2 ESBL-producing isolates	32				CHL, KAN, STR, SUL, TET, TMP			(Politi et al., 2005)
S. Virchow	Healthy/meat	2000-2003	Belgium	13 ESBL-/AmpC-producing Salmonella	2	1			NAL, STR, SUL, TET, TMP			(Bertrand et al., 2006)
E. coli	Healthy	2003	Spain	51 ESBL-producing E. coli	1, 9, 14, 32,	52	2, 5, 12	CMY-2	CHL, CIP, GEN, KAN, NAL, SUL, SXT, TET, TMP, TOB		A/C2, FII, K, N, Q	(Blanc et al., 2006)
E. coli	Meat (pigeon)	2002	China	3.0 (734)	14	1b						(Duan et al., 2006)
S. Bredeney, S. Virchow	Healthy	1999, Unknown	Canada, Ireland	29 S. enterica and 38 E. coli isolates	2			CMY-2	SUL, TET, TMP		A/C, P	(Hopkins et al., 2006)
E. coli	Sick	2003-2004	France	2 ESBL-producing isolates	1	1			CHL, NAL, STR, SUL, TET, TMP			(Meunier et al., 2006)

Species	Sample source	Year	Country	% (Sample #)	CTX-M	TEM	SHV	AmpC and other β-lactamase	Other resistance	MLST	Inc group	Reference
S. Virchow, S. Enteritidis	Healthy	1999-2004	Spain	2.5 (120)	9	1b			NAL, STR, SUL, SXT, TET			(Riano et al., 2006)
S. Infantis	Meat	2004-2005	Japan	6.9 (29)				CMY-2	CHL, KAN, NAL, STR, SXT, TET			(Taguchi et al., 2006)
S. Agona, S. Derby, S. Infantis, S. Paratyphi B, S. Typhimurium	Healthy	2001-2005	Belgium	9 ESBL-producing Salmonella		52			CHL, FLO, STR, SP, SUL, TMP			(Cloeckaert et al., 2007)
E. coli	Healthy	2005	France	11.0 (112)	1	1		AmpC	NAL, RIF, SUL, TET, TMP		11	(Girlich et al., 2007)
E. coli	Healthy/meat	2006	Tunisia	7.7 (78)	1, 8, 14	1	5		CIP, GEN, NAL, STR, SUL, SXT, TET	aadA1, aac(3)-II, dfrA1, sul1, sul2, tet(B), gyrA, parC		(Jouini et al., 2007)
E. coli	Healthy/sick	2003-2005	China	4.6 (389)	14, 27	1b		CMY-2	AMI, APR, FLO, GEN, KAN, NEO			(Liu et al., 2007)
S. Enteritidis	Meat	1999-2005	Japan	2.1 (48)	14				NAL, OFL			(Matsumoto et al., 2007)
S. Typhimurium	Meat	2000-2005	Mexico	3.9 (357)				CMY-2				(Zaidi et al., 2007)
S. Heidelberg	Healthy/meat	2001-2004	Canada	12 isolates				CMY-2	CHL, GEN, KAN, STR, SUL, TET	aadA1, florR, strA, sul1, tet(A), tet(B)	A/C, I1	(Andrysiak et al., 2008)

Table 1. (Continued)

Species	Sample source	Year	Country	% (Sample #)	CTX-M	TEM	SHV	AmpC and other β-lactamase	Other resistance	MLST	Inc group	Reference
S. Braenderup, S. Enteritidis, S. Livingstone	Healthy/meat	2005–2006	Italy	0.6 (2162)			12		GEN, NAL, STR, SUL, TET			(Chiaretto et al., 2008)
Shigella flexneri	Sick	2004	China	Not specified		1, 1d, 116			AMI, GAT, LEV			(Hu et al., 2008)
E. coli	Healthy	1998-2004	Portugal	10.0 (20)		52			NEO, SPE, STR, TET			(Machado et al., 2008)
E. coli, K. pneumoniae	Healthy/meat	1998-2004	Portugal	70.0 (20)	1	52	2		APR, CHL, CIP, GEN, KAN, NAL, NEO, NET, SPE, STR, SUL, TET, TMP, TOB	aadA1, aadA2, dfrA1, dfrA12, estX, sat2		(Machado et al., 2008)
E. coli	Healthy	2007	Belgium	45.0 (295)-ESBL, 43.0 (295)-AmpC	1, 2, 14, 15	52, 106		CMY-2, AmpC	CHL, ENR, GEN, KAN, NAL, NEO, STR, SUL, TET, TMP			(Smet et al., 2008)
E. coli	Meat	2006	UK	13.1 (129)	1, 2, 8, 14							(Warren et al., 2008)
S. Heidelberg	Meat	2002-2006	USA	10.0 (10119)				CMY				(Zhao et al., 2008)
E. coli	Meat	2006	Denmark	1.3 (1650)		1, 1b, 52		CMY-2, AmpC			I1, K, X1	(Bergenholtz et al., 2009)

Species	Sample source	Year	Country	% (Sample #)	CTX-M	TEM	SHV	AmpC and other β-lactamase	Other resistance		MLST	Inc group	Reference
E. coli	Healthy	2004	Portugal	44.7 (76)	14, 32, 14a	52, 1b		AmpC	CHL, CIP, GEN, NAL, STR, SXT, TET, TOB	aadA, tet(A), tet(B), sul1, sul2, sul3, cmlA, aac(3)-I, aac(3)-II, aac(3)-IV, gyrA, parC			(Costa et al., 2009)
S. Agona, S. Kentucky, S. Paratyphi B dT+, S. Typhimurium,	Healthy/ meat	2003-2007	German	9 ESBL-producing isolates	1	20, 52		CMY-2	KAN, NAL, NEO, SPT, STR, SUL, SXT, TET, TMP	aadA1, aadB, aphA1, dfrA1, sat2, strA/B, sul1, sul2, tet(A), tet(B), gyrA			(Rodriguez et al., 2009)
E. coli, K. pneumoniae, S. Infantis, S. Virchow, S. Typhimurium	Healthy	2001, 2004, 2007	Belgium	9 ESBL isolates	2, 15	52			CHL, KAN, NAL, NEO, STR, SUL, TET, TMP			HI2, I1	(Smet et al., 2009)
E. coli	Sick	2007-2008	China	60.8 (51)	14, 24, 65	1, 57			AMI, DOX, ENR, FLO, GEN, SXT				(Yuan et al., 2009)
E. coli	Meat	2007	Tunisia	34.6 (26 food samples)	1			CMY-2, AmpC	CIP, NAL, STR, SUL, SXT, TET	aac(3)-II, aadA, dfrA1, dfrA17, qacEA1, sul1, sul2, sul3, tet(A), tet(B)			(Ben Slama et al., 2010)

Table 1. (Continued)

Species	Sample source	Year	Country	% (Sample #)	CTX-M	TEM	SHV	AmpC and other β-lactamase	Other resistance		MLST	Inc group	Reference
E. coli	Healthy	2007	Italy	67 ESBL-producing E. coli	1, 32	1	12		CHL, CIP, FLO, GEN, NAL, STR, SUL, TET, TMP	sul1, sul2, sul3		FIB, I1, N	(Bortolaia et al., 2010b)
E. coli	Healthy	2009	Italy	33 ESBL-producing E. coli	CTX-M-1, -2, -9 group	1			STR, SUL, TET, TMP			I1, N	(Bortolaia et al., 2010a)
S. Kentucky	Healthy/ broiler house dust samples	2008-2009	Ireland	6.1 (115)			12	CMY-2		aadB, sat			(Boyle et al., 2010)
E. coli	Healthy (ostrich)	Not specified	Portugal	5.6 (54)	14	1b, 52			CHL, CIP, NAL, STR, SXT, TET	aadA1, aadA5, dfrA17, sul1, sul3, tet(A)			(Carneiro et al., 2010)
S. Llandoff	Healthy	2006	France	1 ESBL-producing Salmonella	1				SUL, TET				(Cloeckaert et al., 2010)
E. coli	Healthy	2003	Spain	57 ESBL- and AmpC producing E. coli	1, 9, 14, 14b, 32	52	12	CMY-2			115, 131, 362, 648, 1484		(Cortes et al., 2010)
E. coli	Healthy	2010	Denmark	27.0 (197)			2	CMY-2					(DANMAP, 2010)

Species	Sample source	Year	Country	% (Sample #)	CTX-M	TEM	SHV	AmpC and other β-lactamase	Other resistance	MLST	Inc group	Reference
E. coli	Meat	2008	UK	141 resistant E coli from 62 batches	CTX-M-2, -8 group			AmpC				(Dhanji et al., 2010)
E. coli	Healthy	2006	Netherlands	15.6 (135)	1, 2	52	2	CMY-2	CHL, CIP, GEN, NAL, NEO, STR, SUL, TET, TMP		F, FIB, FIC, HI1, HI2/P, I1, K, P, Y	(Dierikx et al., 2010a)
S. Agona, S. Braenderup, S. Indiana, S. Infantis, S. Parathyphi B var. Java	Healthy	2006	Netherlands	5.8 (359)	1, 2	20, 52		ACC-1	CIP, NAL, NEO, STR, SUL, TET, TMP		ColE, HI1, H2/P, I1, N	(Dierikx et al., 2010a)
E. coli	Meat	2006-2007	Spain	66.7 (20 meat samples)	CTX-M-1, -9 group		12	CMY-4				(Doi et al., 2010)
E. coli	Meat	2006-2007	USA	90.0 (20 meat samples)	CTX-M group 1	1		CMY-2				(Doi et al., 2010)
E. coli	Healthy	2008-2009	Czech Republic	2.9 (304 samples)	1, 14		12	CMY-2	CIP, OFL, TET			(Kolar et al., 2010)
E. coli	Healthy	2006-2007	China	25.0 (224)	3, 14, 15, 24, 55, 64, 65	1			AMI, CHL, CIP, GAT, GEN, NAL, SXT, TET		aadA1, aadA2, aadA5, aadA22, aar-3, dfrI, dfrA12, dfrA17	(Li et al., 2010a)
E. coli	Healthy/sick	1970-2007	China	7.2 (696)	3, 14, 15, 65	52	12					(Li et al., 2010b)

Table 1. (Continued)

Species	Sample source	Year	Country	% (Sample #)	CTX-M	TEM	SHV	AmpC and other β-lactamase	Other resistance	MLST	Inc group	Reference
E. coli	Healthy/sick	2005-2007	Nigeria	11.0 (805)					CHL, CIP, DOX, ERY, GEN, TET			(Mamza et al., 2010)
S, Enteritidis, S. Haardt, S. Indiana, S. Typhimurium	Healthy/environment	2006-2007	Korea	3.3 (91)	1			DHA-1				(Rayamajhi et al., 2010)
S. Infantis	Processing plant	2004-2006	Japan	Not specified		1, 52			STR, SUL, TRI	aadA1, tetA, sul1, dfrA5		(Shahada et al., 2010)
E. coli	Sick	2001-2006	Japan	10.1 (89)	2, 15, 25		2					(Asai et al., 2011)
E. coli	Healthy	2007, 2009	Italy, Denmark	33 ESBL-producing E. coli	1, 2, 9, 32		12		CHL, CIP, FLO, GEN, NAL, STR, SUL, TET, TMP	10, 23, 48, 93, 115, 354, 398, 746, 752, 1137, 1303, 1564, 1626, 1628, 1629, 1630, 1631, 1632, 1633, 1634, 1635, 1636, 1637, 1638	FIB, I1, N	(Bortolaia et al., 2011)
E. coli	Healthy/meat	2004-2006	Japan	11.8 (102)	2, 14		2	CMY-2				(Hiroi et al., 2011)
E. coli	Healthy	2008-2010	China	58.5 (460)	14, 24, 27, 55, 65				AMI, CHL, CIP, GEN, NAL, TET, TRI			(Ho et al., 2011)

Species	Sample source	Year	Country	% (Sample #)	CTX-M	TEM	SHV	AmpC and other β-lactamase	Other resistance	MLST	Inc group	Reference
E. coli	Healthy	Not specified	UK	0.019% (median values for the % of E. coli organisms containing CTX-M gene)	1							(Horton et al., 2011)
Salmonella spp.	Healthy/ meat	2004-2008	Korea	2.2 (46 preselected for anti-microbial resistance)	CTX-M	TEM			aadA2, qnrB, strA, strB, sul2, sul3, tet(A)			(Hur et al., 2011)
E. coli, S. Java	Meat	2010	Netherlands	94.0 (98 retail meat samples)	1, 2	20, 52	2, 12			10, 23, 48, 117, 624, 1403, 1564, 1594, 1901	I1	(Leverstein-van Hall et al., 2011)
E. coli, S. Agona, S. Infantis, S. Java	Healthy	2006	Netherlands	12 (not specified) 35 ESBL	1, 2	20, 52	2			10, 48, 58, 155, 641, 665, 752	I1	(Leverstein-van Hall et al., 2011)
E. coli, K. pneumoniae, other Entero-bacteriaceae	Meat	2009	Netherlands	76.8 E. coli, 7.7 K. pneumoniae, 5.1 other species (isolated from 89 meat samples)	1, 2, 14, 15	52	2, 12					(Overdevest et al., 2011)

Table 1. (Continued)

Species	Sample source	Year	Country	% (Sample #)	CTX-M	TEM	SHV	AmpC and other β-lactamase	Other resistance	MLST	Inc group	Reference
E. coli	Healthy/ meat	2008-2009	UK	3.6 (chicken cecal samples); 54.5 (chicken meat); 6.9 (turkey boot swab); 5.2 (turkey meat)	1, 3, 14, 15, 55	TEM		OXA, CIT-M group			AC, A/C, F, FIA, FIB, I1-γ, K, P	(Randall et al., 2011)
E. coli	Healthy	2010	Sweden	34.0 (200)	1			CMY-2	CIP, KAN, NAL, STR, SUL, TET	qnrB		(SVARM, 2011)
S. Enteritidis, S. Essen	Healthy/ sick/meat	1995-2009	Korea	1.8 (171)-meat, 18.5 (54)-healthy & sick	15				NAL	gyrA		(Tamang et al., 2011)
E. coli	Meat	2009	Denmark	2 (153) [Denmark]; 1.2(173) [Import]	1, 2	20, 52	12	CMY-2, AmpC		4, 10, 57, 88, 155, 156, 359, 371, 1276, 1303, 1494, 1515, 1517, 1518, 1549, 1550, 1551, 1563, 1564		(Agerso et al., 2012)
E. coli	Healthy	2011	Tunisia	58.8 (17)	1	1b, 135		CMY-2	CIP, KAN, NAL, STR, SUL, SXT, TET	aadA1, aadA5, dfrA1, dfrA15, dfrA17 intI1, qacEΔ1, strA-strB, sul1, sul2, tet(A), tet(B)		(Ben Sallem et al., 2012)
E. coli	Meat	2010	Netherlands	94 (98 meat samples)	1, 2	20, 52	2, 12	CMY-2	CIP, TET, TOB	10, 23, 57, 117, 354		(Cohen Stuart et al., 2012)

Species	Sample source	Year	Country	% (Sample #)	CTX-M	TEM	SHV	AmpC and other β-lactamase	Other resistance	MLST	Inc group	Reference
E. coli	Meat	2007, 2010	Spain	62.5 (year 2007), 93.3 (year 2010)	1, 15		12		AMI, CIP, FOS, GEN, NAL, TOB			(Egea et al., 2012)
E. coli	Healthy	2010-2011	Switzerland	63.4 (93)	CTX-M-1 group, 1	TEM, 1, 52	12		CHL, CIP, GEN, NAL, STR, SXT, TET			(Geser et al., 2012)
E. coli	Healthy	2009	Italy	7.9 (101)	1, 2, 14		12			95, 117, 602, 617, 683, 1011, 1818,		(Giufre et al., 2012)
E. coli	Healthy (chicken, common teal, duck)	2009	Bangladesh	30.0 (90)	1, 14 like, 15					131, 206, 224, 405, 448, 648, 744, 1312, 1408, 2141, 2450, 2690, 2691, 2692, 2693		(Hasan et al., 2012)
E. coli, K. pneumoniae	Healthy	2007	Japan	60.0 (30)-broiler (E. coli), 5.9 (17)-layer (E. coli), 3.3 (30)-broiler (K. pneumoniae)	2, 14, 15, 44	1	12					(Hiroi et al., 2012)
E. coli, E. fergusonii, E. cloacae, P. mirabilis, S. fonticola	Meat	2011	Germany	43.9 (399)	1, 2, 65	52	2, 2a, 12		CIP, TET			E. coli, E. fergusonii, E. cloacae, P. mirabilis, S. fonticola

Table 1. (Continued)

Species	Sample source	Year	Country	% (Sample #)	CTX-M	TEM	SHV	AmpC and other β-lactamase	Other resistance	MLST	Inc group	Reference
E. coli	Duck	2006	China	50.4 (230)	14a, 14b, 24b, 24e, 27, 55, 105	1	12	CMY-2, DHA-1	AMI, CHL, CIP, FLO, GEN, TET	8	FII, I1, N	(Ma et al., 2012)
E. coli	Healthy/farm	2010	Tunisia	41.9 (136)-healthy, 66.7 (36)-farm	1, 15	1		CMY-2	GEN, NAL, NET, NOR, SXT, TET, TOB	aac-6'-Ib-cr, qnrB5, qnrS1	F, FIA, FIB, I1, K, N	(Mnif et al., 2012)
E. coli	Meat	2011	USA	7.7 (104)				CMY-2	GEN, TET	117, 131	I1	(Park et al., 2012)
S. Hadar, S. Infantis, S. Manhattan, S. Typhimurium, S. Schwarzengrund	Meat	2006–2011	Japan	8.7 (1252)								(Taguchi et al., 2012)
S. Kentucky	Healthy (turkey)	2010	Poland	1 isolate	CTX-M				CIP, GEN, KAN, NAL, STR, SUL, TET			(Wasyl and Hoszowski, 2012)
K. pneumoniae	Healthy	2009	China	3 isolates	14	1		ACT-type	AMI, CIP, DOX, ENR, FLO, FOS, GEN, LEV, NEO			(Wu et al., 2012)

Species	Sample source	Year	Country	% (Sample #)	CTX-M	TEM	SHV	AmpC and other β-lactamase	Other resistance	MLST	Inc group	Reference
E. coli	Meat	2010	Sweden	44.0 (100)	1	1		CMY-2	CIP, KAN, NAL, STR, SUL, TET	10, 38, 69, 115, 117, 135, 212, 373, 648, 770, 1594, 1640, 2184, 2207, 2370, 2167, 2183	F, FIB, I1, K	(Borjesson et al., 2013)
E. coli	Healthy	2009	Netherlands	At 85% of the farms the prevalence was ≥80%	1, 2	1, 52, 52c	12	CMY-2		aadA1, aadA2, aadA4, catA1, cmlA1, dfrA1, dfrI2, dfrA17, dfrA19, ereB, floR, sul1, sul2, sul3, strA, strB, tet(A)	38, 70, 93, 115, 359, 380, 420, 602, 665, 770, 789, 997, 1358, 2309	B/O, ColE, FIB, FIC, Frep, I1, N, K	(Dierikx et al., 2013a)
E. coli	Healthy/environment	2009-2010	Netherlands	8.7 (957)	1, 2	20, 52, 52c	12	CMY-2, AmpC	CHL, CIP, GEN, KAN, NAL, STR, SUL, TET, TMP			(Dierikx et al., 2013b)
E. coli	Healthy	2011-2012	Tunisia	4.1 (193)	1, 9				APR, ENR, KAN, NAL, STR, SUL, TET, TOB, TRI		B/O, F, FIA, FIB, I1, P	(Grami et al., 2013)
E. coli	Healthy	2004-2009	Japan	5.2 (1377)	2, 14, 25	1	1, 2, 2a, 5, 12	CMY-2, AmpC	CHL, CIP, GEN, KAN, NAL, SXT, TET		A/C, B/O, I1, Iγ	(Hiki et al., 2013)
E. coli	Healthy	2011	Japan	44.5 (164)	1, 2, 55							(Kameyama et al., 2013)

Table 1. (Continued)

Species	Sample source	Year	Country	% (Sample #)	CTX-M	TEM	SHV	AmpC and other β-lactamase	Other resistance	MLST	Inc group	Reference
E. coli	Healthy	Not specified	Germany	67.1 (420)	CTX-M	1, 52	12	CMY-2				(Laube et al., 2013)
E. coli	Sick (chicken, duck)	2003-2007	China	30.0 (247)	14, 27				CHL, FLO, GEN, KAN, TET, NAL, DOX, ENR, CIP, LEV		F2:A-:B-, F33:A-:B-, N	(Liao et al., 2013)
E. coli, Enterobacter cloacae, Proteus mirabilis	Healthy	2010	Germany	ESBL producers were found in 88.6% and 72.5% of carcasses and ceca; AmpC producers were found in 52.9% and 56.9% of carcasses and ceca	CTX-M			AmpC				(Reich et al., 2013)
E. coli	Healthy	Not specified	Germany	1.8 (438)					DOX, GEN, NEO, SPE, STR, SXT, TOB			(Schwaiger et al., 2013)
S. Agona, S. Schwarzengrund	Healthy/ meat/ environment	2008-2009	Brazil	14.0 (93)	2				CIP, ENR, NAL, STR, SXT, TET			(Silva et al., 2013)

Species	Sample source	Year	Country	% (Sample #)	CTX-M	TEM	SHV	AmpC and other β-lactamase	Other resistance	MLST	Inc group	Reference
E. coli	Meat	2009	Netherlands	94.0 (98)				CMY-2	CHL, CIP, TET	23, 38, 93, 115, 117	I1, K	(Voets et al., 2013)
S. Abony, S. Agona, S. Edinburg, S. Enteritidis, S. Gueuletapee, S. Haardt, S. Heidelberg, S. Indiana, S. Shubra, S. Thompson, S. Typhimurium, S. Uppsala	Meat	2011	China	8.6 (699)								(Wu et al., 2013)
E. coli, K. pneumoniae	Healthy	2012-2013	Spain	91.1 (260)-E. coli, 1.2 (260)-K. pneumoniae	CTX-M-1, -2, -9 group				CIP, GEN, NAL, SXT			(Abreu et al., 2014)
E. coli	Healthy/meat	2009-2011	Denmark	93% parent farms, 27% broiler flocks, 6.5% meat	1		2	CMY-2	NAL, NEO, SUL, TET	10, 23, 38, 48, 69, 88, 115, 131, 206, 212, 219, 350, 410, 616, 746, 1056, 1303, 1518, 1585, 1594, 1775, 3272	I1, K	(Agerso et al., 2014)

Table 1. (Continued)

Species	Sample source	Year	Country	% (Sample #)	CTX-M	TEM	SHV	AmpC and other β-lactamase	Other resistance	MLST	Inc group	Reference
E. coli	Healthy	2010-2011	Tunisia	10 isolates	1			CMY-2	CIP, NAL, STR, SUL, SXT, TET	sul2, tet(A)	10, 57, 88, 117, 155, 2016, 2164, 2255, 3632	B/O, F4:A-:B1, F18:A-:B1, F18:A5:B1, F24:A-:B1, F45:A-:B1, F55:A-:B1, F55:A1:B27, II, K, Y	(Ben Sallem et al., 2014)
E. coli	Flies at Poultry Farms	2011	Netherlands	10.5 (flies), 82.0 (manure & rinse water in broiler farm), 80.0 (manure & rinse water in layer farm)	1	52	12					(Blaak et al., 2014)
E. coli	Meat	2009-2011	Denmark	83.8 (Not specified)	1, 2	20, 52	2a, 12	CMY-2, AmpC				(Carmo et al., 2014)
S. Agona, S. Enteritidis, S. Indiana, S. Infantis, S. Minnesota, S. O4:-, S. Paratyphi B, S. Rissen, S. Typhimurium, S. Virchow	Healthy	2008-2011	Belgium	13.1 (452)	1, 2, 9	52	12	CMY-2	CHL, NAL, STR, SXT, TET			(de Jong et al., 2014)

Species	Sample source	Year	Country	% (Sample #)	CTX-M	TEM	SHV	AmpC and other β-lactamase	Other resistance		MLST	Inc group	Reference
E. coli	Meat	2010-2011	Sweden	15.0-95.0 (Not specified)	1, 2, 8, 25	1, 19, 52, 135	12	CMY-2	CHL, CIP, FLO, GEN, KAN, NAL, STR, SUL, TET, TRI			A/C, HI2, I1, K, O, P	(Egervarn et al., 2014)
E. coli	Healthy	2011-2012	Brazil	19 isolates	2				CHL, CIP, GENM LEV, NAL, SXT, TET	qnrB19	93, 155, 2309	B/O, F, FIB, I1, K	(Ferreira et al., 2014)
E. coli	Healthy/ nearby river water	2012-2013	China	11.2 (258)	CTX-M	TEM			CHL, CIP, GEN, KAN, NAL, NOR, STR, SXT, TET				(Gao et al., 2014)
E. coli	Healthy (chicken, duck, geese)	2010-2012	China	2.1 (659)	55	1		CMY-2, CMY-41, CMY-64		aac-(6')-1b-cr, aadA1, aadA2, aadA5, aadA22, dfrA1, dfrA17, floR, oqxA, qnrB6, qnrS1, rmtB, sat1	48, 69, 155, 156, 354, 362, 648, 1431, 2294, 2690, 3245	A/C, FIB, HI2, K	(Guo et al., 2014)
E. coli	Healthy (pigeons)	2011	Bangladesh	4.7 (150)	15				NAL, SXT		1408, 3489, 3490, 3491, 3492		(Hasan et al., 2014)
E. coli	Healthy	2010-2011	Netherlands	Not specified	1, 2, 9, 14, 32, 52	1b, 52	12	CMY-2			10, 38, 57, 88, 117, 189, 351, 355, 373, 641, 648, 1146, 1640, 1818, 1775, 2223, 2509	A/C, B/O, FIB, FII, HI2, I1, K, N, XI	(Huijbers et al., 2014)

Table 1. (Continued)

Species	Sample source	Year	Country	% (Sample #)	CTX-M	TEM	SHV	AmpC and other β-lactamase	Other resistance	Other resistance	MLST	Inc group	Reference
E. coli	Meat	2010	Japan	50.0 (68)	1, 2, 3, 8, 15	52, 135	2, 12		CHL, CIP, GEN, LEV			F, FIB, I1-γ, N, P	(Kawamura et al., 2014)
E. coli	Healthy/ farm & surrounding area	Not specified	Germany	Not specified	CTX-M	1, 52	12	CMY-2					(Laube et al., 2014)
E. coli	Healthy	2012	China	49.5 (202)	CTX-M	1, 206				aac(69)-Ib-cr, qnrS			(Li et al., 2014)
S. Enteritidis, S. Heidelberg, S. Infantis, S. Manhattan, S. Schwarzengrund	Meat	2005-2012	Japan	4.6 (524)	2, 14	1, 20v, 52							(Matsumoto et al., 2014)
E. coli	Meat	2011-2012	Gabon	23.3 (60)	1, 14, 15, 27, 32	TEM	SHV		CIP, GEN				(Schaumburg et al., 2014)
E. coli	Meat	2013	Switzerland	73.3 (75)	1	1	12	CMY-2, CMY-4	CHL, CIP, KAN, NAL, SUL, TET, TRI	aadA1-like, aadA2-like, aadA4-like, catB3 like, cmlA1-like, dfrA7, dfrA17, dfrA19, ereB, IntI1, strAB, sul1, sul2, sul3, tet(A), tet(B)	10, 38, 48, 57, 69, 93, 155, 371, 752, 770, 1056, 1158, 1304, 1564, 1850, 2248, 2307, 4007	B/O, FIB, FII, HI1, I1, I2, K, Y	(Vogt et al., 2014)

Species	Sample source	Year	Country	% (Sample #)	CTX-M	TEM	SHV	AmpC and other β-lactamase	Other resistance	MLST	Inc group	Reference
E.coli, Enterobacter spp, Klebsiella oxytoca, Klebsiella pneumoniae	Meat	2013	Egypt	65.1 (106)	14, 15	TEM	SHV					(Abdallah et al., 2015)
E. coli	Poultry Farm Environment	2011-2012	Netherlands	1,107 E. coli	1, 2	52	12		CHL, CIP, NAL, STR, SUL, TET, TRI	10, 57, 58, 117, 155, 189, 212, 219, 295, 371, 420, 448, 616, 683, 997, 1158, 1564, 1594, 1610, 1684, 1818, 1844, 2079, 2309, 3519, 4980, 4994		(Blaak et al., 2015)
S. Bareilly, S. Enteritidis, S. Infantis, S. Richmond, S. Virchow	Meat	2014	Korea	29.0 (100)	1, 15							(Choi et al., 2015)
S. Bellevue, S. Enteritidis	Meat	2014	Korea	54.5 (11)	15				AMI, NAL, STR, TET			(Chon et al., 2015)
E. coli	Meat	2015	Denmark	36.1 (545)	1	52b, 84	2, 12	CMY-2, AmpC				(DANMAP, 2015)
S. Infantis	Healthy/ meat	2011-2014	Italy	29 extended-spectrum cephalosporin resistant isolates	1				CHL, CIP, GEN, KAN, NAL, STR, SUL, TET, TRI	aac(6')-ly, aadA1, aph(3')-Ic, dfrA1, dfrA14, sul1, tet(A)	P	(Franco et al., 2015)

Table 1. (Continued)

Species	Sample source	Year	Country	% (Sample #)	CTX-M	TEM	SHV	AmpC and other β-lactamase	Other resistance		MLST	Inc group	Reference
S. Infantis	Healthy/meat	2011-2014	Italy	29 extended-spectrum cephalosporin resistant isolates	1				CHL, CIP, GEN, KAN, NAL, STR, SUL, TET, TRI	aac(6')-ly, aadA1, aph(3')-Ic, dfrA1, dfrA14, sul1, tet(A)		P	(Franco et al., 2015)
E. coli	Meat	2013	Italy	82.2 (163)	CTX-M-1, -2, -9 group	TEM	SHV	CIT-like	CIP	qnrA	131		(Ghodousi et al., 2015)
E. coli	Healthy	2011-2012	Netherlands	49 isolates	1	1b, 52		CMY-2				I1, K	(Huijbers et al., 2015)
E. coli	Healthy	2009	Denmark	3 CTX-M-1-positive E. coli	1							HI2/P, I1	(Jakobsen et al., 2015)
E. coli	Healthy/sick	2013-2014	India	6.34 (252)	CTX-M	TEM	SHV	AmpC	AMI, CHL, CIP, GAT, NOR, LEV, LOM, TET, TOB	int1, qnrB, sul1			(Kar et al., 2015)
E. coli	Healthy	2013	Tunisia	26.2 (65)	1	1			NAL, NOR, STR, SXT, TET	qnrB1, qnrS1, sul1, sul2, sul3, tet(A), tet(B)			(Kilani et al., 2015)
Salmonella isolates	Meat/environment in farm & slaughter-house	2012-2013	Korea	23.3 (160)	15				GEN, NAL, STR, TET			FIIS, FIA, FIB, I1, HI2, K/B, P	(Kim et al., 2015)

Species	Sample source	Year	Country	% (Sample #)	CTX-M	TEM	SHV	AmpC and other β-lactamase	Other resistance	MLST	Inc group	Reference
E. coli	Meat	Not specified	Brazil	42.1 (121)	CTX-M-1, -2, -8 group		SHV	CIT	CHL, CIP, ENR, GEN, NAL, NOR, SXT, TET			(Koga et al., 2015b)
E. coli	Meat	2013	Brazil	32.2 (121)	CTX-M-1, -2, -8 group							(Koga et al., 2015a)
E. coli	Healthy/meat	2012	Korea	39.7 (156)	1	TEM	SHV		CIP, GEN	10, 38, 43, 57, 93, 95, 101, 117, 155, 162, 208, 354, 449, 457, 602, 1056, 1286, 1403, 1431, 1463, 2309, 2667, 2847, 3189		(Lim et al., 2015)
E. coli	Healthy	Not specified	Algeria	17 ESBL-producing E. coli isolates	CTX-M	TEM						(Mezhoud et al., 2015)
E. coli	Healthy	Not specified	Algeria	17 ESBL-producing E. coli isolates	CTX-M	TEM						(Mezhoud et al., 2015)
E. coli	Meat	2012-2013	Germany	81.6 (?)	1, 14	52c	12	CMY-2				(Pacholewicz et al., 2015)
E. coli	Healthy/meat	2013	Ghana	20.9 (153)	1, 15, 61			AmpC, CIT		10, 38, 117, 124, 155, 156, 162, 205, 212, 354, 542, 642, 1158, 1304, 1431, 2167, 2461, 4028, 4120, 4121, 4122		(Rasmussen et al., 2015)

Table 1. (Continued)

Species	Sample source	Year	Country	% (Sample #)	CTX-M	TEM	SHV	AmpC and other β-lactamase	Other resistance	MLST	Inc group	Reference
E. coli	Healthy	2011	India	19.2 (120)								(Samanta et al., 2015a)
E. coli	Flies in Broiler Farm	2012	Spain	6.2 (682)	1, 9, 14				CHL, GEN, KAN, STR, SUL, TET, TRI	qnrS		FIA, FIB, HI1, HI2, I1, K, N, P	(Sola-Gines et al., 2015)
E. coli	Healthy/ sick/meat	2011-2013	China	97.9 (195)	3, 13, 15, 55, 14, 64, 65, 79, 101, 123, 132	1	5		CHL, CIP, GEN, LEV, SXT, TET			(Tong et al., 2015)
E. coli	Meat, contact surfaces of weighing scales, cutting boards	2012-2013	Malaysia	48.8 (240)								(Aliyu et al., 2016)
E. coli	Healthy	2014	Algeria	32.8 (61)	1	1	12	CMY-2	AMI, CIP, KAN, NAL, STR, TOB	38, 744, 1011, 2179, 5086		aadA, qnrA	(Belmahdi et al., 2016)
E. coli	Healthy	Not specified	Zambia	20.1 (384)	CTX-M	TEM	SHV		CHL, CIP, GEN, NAL, NOR, STR, SXT, TET			(Chishimba et al., 2016)

Species	Sample source	Year	Country	% (Sample #)	CTX-M	TEM	SHV	AmpC and other β-lactamase	Other resistance	MLST	Inc group	Reference
E. coli, Escherichia fergusonii, K. pneumoniae	Healthy	2011-2012	Brazil	16 isolates	2, 8, 15				CHL, CIP, GEN, LEV, NAL, SXT, TET		A/C, F, FIB, FIC, HI1, I1, K, Y	(Ferreira et al., 2016)
E. coli	Healthy/meat	2006-2012	Denmark	58 CMY-2 isolates				CMY-2		10, 23, 38, 48, 69, 88, 93, 115, 117, 131, 206, 212, 219, 350, 410, 428, 616, 745, 746, 919, 1056, 1196, 1303, 1518, 1585, 1594, 1640, 1775, 2040, 3272, 4048, 4124, 4125, 4240, 4243	A/C, FII, I1-Iγ, K, ND1, ND2, R	(Hansen et al., 2016)
Aeromonas caviae	Meat	Not specified	China	5.9 (34)				PER-3	ant(3'')-Ij– aac(6')-Ib, catB3, qacEΔ1, sul1			(Li et al., 2016)
E. coli, K. pneumonia, P. mirabilis	Meat	2015	Singapore	56 ESBL isolates	1, 2, 8, 9							(Lim et al., 2016)
E. coli	Healthy	2013	Tunisia	35.0 (137)	1, 14, 15	1		CMY-2	AMI, CHL, CIP, GEN, MIN, NAL, SXT, TET, TOB	aadA1, aadA2, dfrA1, dfrA12, dfrA14, sat2, sul1, sul3, tetA, tetB	58, 93, 117, 155, 212, 349, 350, 405, 542, 1096, 1196, 1431, 2197, 4968	F, FIA, FIB, I1, K, N, P, Y	(Maamar et al., 2016)

Table 1. (Continued)

Species	Sample source	Year	Country	% (Sample #)	CTX-M	TEM	SHV	AmpC and other β-lactamase	Other resistance	MLST	Inc group	Reference
E. cloacae, E. fergusonii	Egg	Not specified	Belgium	3.2 (186)		1, 52	12	ACT-39	APR, FLO, GEN, NEO, SUL, TET, TRI			(Mezhoud et al., 2016)
E. coli	Meat	2012-2014	Vietnam	45.5 (150)	CTX-M-1, -2, -8, -9, -25 group	1, 135	12	CIT, DHA groups	CHL, CIP, FOS, GEN, KAN, NAL, STR, SXT, TET			(Nguyen do et al., 2016)
E. coli	Healthy/ meat	2009-2014	Nigeria	1.0 (405)	15	1		OXA-1, AmpC	CHL, ENR, FLO, GEN, KAN, NAL, STR, SUL, TET, TMP	10, 405	FIA, FIB, I1	(Ojo et al., 2016)
Salmonella spp	Processing plant	Not specified	Brazil	45.0 (98)					AMI, CIP, ENR, GEN, NAL, NEO, STR, SXT, TET, TOB			(Ziech et al., 2016)
E. coli	Meat	2015	Switzerland	41.3 (80)	1, 2, 8, 14	52	12					(Zogg et al., 2016)

Note: aac(3)-IIa, aac(6')-1b-cr, aadA2, dfrA5, dfrA12, floR, strA, strB, sul1, sul2, tet(A), tet(B), gyrA, parC (E. coli Nigeria row, Other resistance); mcr-1 (E. coli Switzerland row, Other resistance).

Species	Sample source	Year	Country	% (Sample #)	CTX-M	TEM	SHV	AmpC and other β-lactamase	Other resistance	MLST	Inc group	Reference
E. coli, K. pneumoniae, Escherichia fergusonii, Proteus mirabilis, Escherichia Hermannii, Pseudomonas aeruginosa, Bordetella trematum, Acinetobacter spp.	Healthy	2014	India	66.1 (510)								(Brower et al., 2017)
E. coli	Healthy /meat	2008-2013	Colombia	22.6 (541)	2, 8, 15		5, 12	CMY-2		10, 23, 38, 48, 57, 101, 135, 212, 155, 162, 189, 201, 224, 359, 226, 366, 533, 624, 641, 746, 973, 1049, 1158, 1266, 1775, 2040, 2847, 3107, 3910, 4243, 5416	A/C, B/O, HI2, I1, I1-F, K	(Castellanos et al., 2017)
C. freundii, E. coli, K. pneumoniae	Healthy / environment	2014-2016	Germany	20.7 (1133)	1		2, 12	CMY-2		2307		C. freundii, E. coli, K. pneumoniae

Table 1. (Continued)

Species	Sample source	Year	Country	% (Sample #)	CTX-M	TEM	SHV	AmpC and other β-lactamase	Other resistance	MLST	Inc group	Reference
S. Heidelberg, S. Newport	Healthy/meat/environment in farm & slaughterhouse	2011, 2013	Algeria	1.5 (1194)	1	TEM					15, 198	
E. coli	Sick	2012	Egypt	6.0 (50)	14	57	12					(El-Shazly et al., 2017)
S. Infantis	Meat	2010, 2013, 2015	Switzerland	0.2 (496)	65				CHL, GEN, KAN, NAL, STR, SUL, TET, TMP	aac3IVa, aadA1, aph3'-Ic, aph4-Ia, dfrA14, floR, fosA, sul1, tetA	32	(Hindermann et al., 2017)
Salmonella spp	Meat	2015	Japan	2.7 (74)		52		CMY-2	STR, TET			(Kataoka et al., 2017)
Klebsiella spp	Healthy (chicken, kuroiler)	2014-2015	India	10.7 (307)	9	1	12					(Mahanti et al., 2017)
E. coli	Healthy	2013-2014	Vietnam	24 ESBL isolates	9	TEM		AmpC	CHL, CIP, FOS, GEN, KAN, NAL, STR, SXT, TET			(Nakayama et al., 2017)
Enterobacteria	Healthy/environment	2014-2015	Germany	3.7 (909)	1	1, 15, 52	1, 2	CMY-2				(Projahn et al., 2017)

Species	Sample source	Year	Country	% (Sample #)	CTX-M	TEM	SHV	AmpC and other β-lactamase	Other resistance	MLST	Inc group	Reference
E. coli	Sick	2011-2015	Korea	6.7 (75)	14, 15			CMY-2, CMY-6	CIP, GEN, NAL, SXT, TET		FIB, Frep, I1	(Shin et al., 2017)
S. Infantis	Healthy/ meat	2014-2015	USA	5 blaCTX-M-65-positive S. Infantis	65				CHL, GEN, NAL, STR, SXT, SUL, TET	aac(3)-IVa, aadA1, aph(3')-Ic, aph(4)-Ia, dfrA14, floR, fosA3, sul1, tetA, gyrA		(Tate et al., 2017)
E. coli	Meat	2009-2012	Korea	0.8 (1771)	1, 2, 94	1		OXA-1, AmpC	CHL, CIP, GEN, NAL, KAN, STR, TET	117, 155, 350, 359, 457, 533, 1266, 1771, 2309 2491	B/O, FIA, FIB, FrepB, HI1, I1-Iγ, P, Y	(Kim et al., 2018)

*AMI, amikacin; APR, apramycin; CHL, chloramphenicol; CIP, ciprofloxacin; DOX, doxycycline; ENR, enrofloxacin; FLO, florfenicol; FOS, Fosfomycin; GAT, gatifloxacin; GEN, gentamicin; KAN, kanamycin; LEV, levofloxacin; MIN, minocyclin; NAL, nalidixic acid; NET, netilmicin; NEO, neomycin; OFL, ofloxacin; SMX, RIF, rifampin; sulfamethoxazole; SPE, spectinomycin; STR, streptomycin; SUL, sulphonamides; SXT, sulfamethoxazole/trimethoprim; TET, tetracycline; TMP, trimethoprim; TOB, tobramycin.

**Underline in AmpC means AmpC mutation.

***Underline in gyrA, gyrB, parC, parE means floroquinolone mutation.

The prevalence rates of ESBL-producing *Salmonella* recovered from poultry is generally lower than ESBL-producing *E. coli*. Between 1999-2004, frequency of ESBL-positive *S. enterica* isolates in chicken feces in Spain was 2.5% (3/120) (Riano et al., 2006). Prevalence of ESBL/AmpC-containing bacteria in chickens tends to be higher than other food producing animals. Kojima et al. found that layer and broiler chickens (1.7%, 18/1,082) in Japan were contaminated with ESBL/AmpC producers to a higher extent than pigs (0/793) and cattle (0/872) (Kojima et al., 2005). In Portugal, 10% and 5.7% of Enterobacteriaceae isolates recovered from feces of healthy chickens and pigs, respectively, were ESBL producers (Machado et al., 2008). Fecal carriage rate of ESBL-carrying *E. coli* was higher in chickens (25.0%) than in pigs (0%) in China (Li et al., 2010a). de Jong et al. reported that 13.1% (59/452) of *S. enterica* isolates from broiler chickens and 1.9% (7/368) from pigs in Belgium contained the ESBL genes (de Jong et al., 2014). Wasyl et al. isolated cefotaxime-resistant *E. coli* from 33.3% of samples in pigs, 42.3% in layers, 54.5% in broilers and 48.0% in turkey in Poland (Wasyl et al., 2012).

Poultry ESBL/AmpC-Producers Other Than *E. coli* and *Salmonella*

Other ESBL/AmpC-producing bacterial isolates (*Acinetobacter* spp., *Aeromonas caviae, Bordetella trematum, Citrobacter freundii, Enterobacter cloacae, Escherichia fergusonii, Escherichia hermannii, Klebsiella pneumoniae, Proteus mirabilis, Pseudomonas aeruginosa, Serratia fonticola, Shigella flexneri*) from poultry were also documented (Abreu et al., 2014; Brower et al., 2017; Gundogan et al., 2011; Hiroi et al., 2012; Mahanti et al., 2017). About seventy percent of 510 ESBL producers from chicken cloacae were gram-negative bacteria (*Acinetobacter* spp., *Bordetella trematum, E. coli, E. fergusonii, E. hermannii, K. pneumoniae, P. mirabilis, P. aeruginosa*) in India (Table 1) (Brower et al., 2017). On the other hand, ESBL-expressing *Klebsiella* spp. in healthy chickens were documented at low frequency (1.2-3.3%) in Spain and Japan (Abreu et al., 2014; Hiroi et al., 2012).

ESBL/AmpC-Associated Genes in Poultry

An examination of the ESBL/AmpC bacteria in domestic chicken meat samples harboring the most popular β-lactamase genes (CTX-M-2, CTX-M-15, CMY-2 and SHV-12) revealed a country-specific distribution pattern of these genes. In Japan, the most popular CTX-M type from *E. coli* isolates in chicken meat was CTX-M-2 (42.3%) (Table 1) (Kawamura et al., 2014). In Korea, ESBL-expressing *E. coli* isolates from chicken carcasses and feces were found to have CTX-M group 1 (40.3%) (Lim et al., 2015). All ESBL-bearing bacteria from chicken meat in Korea carried CTX-M-15 (Chon et al., 2015; Kim et al., 2015). *E. coli* isolates from poultry in China carried CTX-M (19.8%), TEM-1 (24.3%), and TEM-206 (11.9%) from healthy chicken (Li et al., 2014). Another group found that CTX-M-producing *E. coli* from poultry carried CTX-M-15 (37.2%), CTX-M-65 (22.4%), CTX-M-55 (19.1%), and CTX-M-14 (17.5%) (Tong et al., 2015). Extended-spectrum cephalosporin-resistant *E. coli* recovered from all rectal samples from healthy broiler chickens in Japan contained CMY-2 (66.0%), CTX-M-1 (26.0%) and CTX-M-55 (10.0%) (Kameyama et al., 2013). In Europe, 2.0-70.0% of ESBL/AmpC-producing *E. coli* isolates harboring CTX-M-1 enzyme were recovered from healthy chickens (Abgottspon et al., 2014; Endimiani et al., 2012c; Franco et al., 2015; Geser et al., 2012; Girlich et al., 2007; Leverstein-van Hall et al., 2011; Smet et al., 2008; Vogt et al., 2014; Zurfluh et al., 2014). CTX-M-14, CMY-2 and SHV-12 genes from ESBL-harboring *E. coli* strains (1.7%) were isolated in food-producing animals in Spain (Brinas et al., 2003). Two years later, the same research group reported an increase of β-lactamase genes from 1.7 to 3.4% in *E. coli* isolates from livestock including chicken (Brinas et al., 2005). In the south of Spain, the most popular ESBL type in *E. coli* in chicken meat was SHV-12 (Egea et al., 2012). The detection rate of CTX-M-1 in chicken was reported to be 11.0-78.0% in France, Belgium, Switzerland and UK (Geser et al., 2012;

Girlich et al., 2007; Randall et al., 2011; Smet et al., 2008). In Canada, CMY-2, TEM, and SHV were detected from 26.9%, 17.6%, and 2.1%, respectively, in *E. coli* isolates from chicken meat (Sheikh et al., 2012). *E. coli* carrying CMY-2 and CTX-M-1 were present in 53.7% and 6.9% of the US chicken meat samples (Mollenkopf et al., 2014). In South America, 22.6% (122/541) of the ESBL/AmpC-bearing *E. coli* in poultry production chain commonly contained CMY-2 gene in 75.4% of the isolates followed by SHV-12 gene in 11.5% of the isolates (Castellanos et al., 2017). *E. coli* isolates from chicken in Brazil harbored CTX-M-2 (37.5-100%) and CTX-M-8 (56.3%) genes (Ferreira et al., 2014; Ferreira et al., 2016). In Africa, CTX-M-1, -14, and -15, SHV, TEM and CMY-2 have been reported in healthy chickens (Maamar et al., 2016). Seventy percent (14/20) of ESBL-producing *E. coli* strains from Algeria contained both SHV-12 and TEM-1 (Belmahdi et al., 2016). In Zambia, 384 swab specimens of market-ready chickens indicated the presence of ESBL-producing *E. coli* in 20.1% isolates (Chishimba et al., 2016).

The first report of ESBL-carrying *S. enterica* (serovar 35:c:1,2) in chicken meat product appeared in 2001 from Senegal (Cardinale et al., 2001). The *S. enterica* isolates obtained from chicken feces in Spain contained TEM-1b + CTX-M-9 gene combination. In 2005 and 2006, 0.38% (3/789) and 0.86% (9/1,053) of *Salmonella* from Italy were cephalosporin-resistant and all of them contained SHV-12 (Chiaretto et al., 2008). ESBL-resistant *Salmonella* from broiler chickens in the Netherlands were found to produce various types of β-lactamase genes, such as CTX-M-2, CTX-M-3, CTX-M-28, SHV-2, SHV-12, TEM-1, and TEM-52, and the most popular type was TEM-52 (Hasman et al., 2005). In Belgium, ESBL-containing *Salmonella* from poultry carried CTX-M-2 and TEM-52 genes (Bertrand et al., 2006; Cloeckaert et al., 2007). In France, CTX-M-9 gene was detected in 3.2% of *S.* Virchow isolates from poultry (Weill et al., 2004). Notably, the CTX-M-65 gene that was prevalently found in Chinese *S.* serovar Infantis isolates was also recovered from chicken meat in Switzerland (Bai et al., 2016; Hindermann et al., 2017). TEM-52-carrying *Salmonella* from poultry have also been reported in Japan (Shahada et al., 2010). In North America, 50.0% and 47.0% of ampicillin-resistant *Salmonella* from US retail meat including chicken were CMY and TEM-1 producers, respectively (Zhao et al., 2009). Furthermore, all ceftriaxone- and ceftiofur-resistant *E. coli* or *Salmonella* from retail meat in USA harbored CMY but not CTX-M genes (Park et al., 2012; Zhao et al., 2012; Zhao et al., 2008). *Salmonella* isolated from human clinical samples exhibited a high resistance to fluoroquinolones and cephalosporins in Asian countries (Lee et al., 2009; Van et al., 2012). These results are considered to be related to the high resistance of chicken meat isolates to ciprofloxacin (22.5%), ceftriaxone (12.5-23.4%), and ceftazidime (26.6%) in China and Korea (Yang et al., 2014; Yoon et al., 2014).

In Belgium, ESBL-producing *E. coli* were recovered most frequently in pigs (63.6%, 136/214) than in chickens (58.5%, 269/460) and cattle (32.9%, 69/210) (Ho et al., 2011). Additionally, CTX-M-containing *E. coli* were recovered from pigs (2.0%, 6/300), cattle (3.1%, 3/96), and pigeons (3.0%, 1/33) in slaughterhouses or wholesale markets in Hong Kong whereas none of them were found from chickens, ducks and geese (Duan et al., 2006). Among chicken, pork, and beef, the highest numbers of ESBL producers were found in chicken and most of them were imported from Brazil. The predominant ESBL genotype was CTX-M. However, high percentage (32.6%) of cefazolin-resistant *E. coli* containing SHV-2, CTX-M-25 and CTX-M-15 was recovered from sick food-producing animals in Japan (Asai et al., 2011).

Multidrug-Resistance in ESBL/AmpC-Producers in Poultry

Multidrug-resistant genes conferring resistance to aminoglycosides (*aac(3)-I, aac(3)-II, aac(3)-IIa, aac(3)-IV, aac(3)-IVa, aac(6')-Ib-cr, aac(6')-ly, aadA, aadA1, aadA2, aadA4, aadA5, aadA22, aadB, ant(3')-Ij– aac(6')-Ib, aphA1, aph(3')-Ic, aph(4)-la, strA, strB*), chloramphenicol (*catA1, catB3, cmlA, floR*), quaternary ammonium compound (*qacEΔ1, sul1, sul2, sul3*), tetracycline [*tet(A), tet(B)*] and trimethoprim (*dfrA1, dfrA5, dfrA12, dfrA14, dfrA16, dfrA17, dfrA19*), have been detected in ESBL producers (Table 1) (Boyle et al., 2010; Dierikx et al., 2013a; Doi et al.,

2010; Hindermann et al., 2017; Li et al., 2010a; Maamar et al., 2016; Machado et al., 2008; Tate et al., 2017). Some unique antibiotic-resistant genes, such as fosfomycin (*fosA, fosA3*), macrolide (*ereB*), rifampicin (*aar-3*), and streptothricin (*estX, sat, sat2*), have also been documented in ESBL producers (Boyle et al., 2010; Costa et al., 2009; Dierikx et al., 2013a; Doi et al., 2010; Hindermann et al., 2017; Jouini et al., 2007; Li et al., 2010a; Maamar et al., 2016; Machado et al., 2008; Ojo et al., 2016; Tate et al., 2017). Fosfomycin resistance genes *fosA* and *fosA3* were recently found in CTX-M-65-containing *S.* Infantis isolated from chicken meat and ceca in USA and Switzerland (Hindermann et al., 2017; Tate et al., 2017). Streptothricin-resistant gene *estX* was present on a class 2 integron found among TEM-52-producing Enterobacteriaceae from chickens in Portugal. Other genes, namely *sat* and *sat2* along with CTX-M, SHV, and TEM genes were found in *S.* Kentucky and *S.* Paratyphi B dT+ from healthy chicken and chicken meat in Germany, Ireland, Portugal, and Tunisia (Boyle et al., 2010; Doi et al., 2010; Maamar et al., 2016; Machado et al., 2008). *E. coli* isolates carrying CTX-M, CMY, SHV, and TEM from broilers in the Netherlands were also reported to possess *ereB* gene (Dierikx et al., 2013a). CTX-M-65 and TEM-1-positive *E. coli* isolates from healthy chicken in China harbored *aar-3* (Li et al., 2010a).

Integron gene cassettes in fluoroquinolone-resistant clinical *E. coli* isolates in China contained not only *qnrA* but also OXA-30 β-lactamase and other resistance genes [*aac(6')-Ib, ampR, arr3, catB3, qacEΔ1* and *sul1*] (Robicsek et al., 2006). Different types of integrons [class I: XXII$_1$ (*aadA13-estX*), XXIII$_1$ (*dfrA14-aadA1-catB2*), class II: type IV$_2$ (*aadA1*)] were detected among Enterobacteriaceae from food-producing animals in Portugal (Machado et al., 2008). Plasmid-mediated quinolone resistance determinants, such as *qnrA*, *qnrB*, and *qnrS*, have been discovered from ESBL producers (Costa et al., 2009; Doi et al., 2010; Ghodousi et al., 2015; Hur et al., 2011; Jouini et al., 2007; Kar et al., 2015; Ojo et al., 2016; Randall et al., 2011; Tate et al., 2017). Fluoroquinolone-resistant and ESBL/AmpC-positive *E. coli* isolates in Italy possessed only *qnrA* (Ghodousi et al., 2015). In addition, *qnrB* was discovered in CTX-M, OXA-positive *E. coli* and *S. enteritidis* in Asia (India, Korea) and Europe (UK) (Hur et al., 2011; Kar et al., 2015; Randall et al., 2011). One study indicated that *qnrS* coexisted with TEM-206 and *aac(6')-1b-cr* in the same *E. coli* isolate from a chicken farm in China (Li et al., 2014). CMY-2, CTX-M, SHV, and TEM-containing *E. coli* and *S.* Agona, Infantis, Paratyphi B dT+ and Virchow isolates from chicken, goulash, and turkey in Germany, Nigeria, Portugal, Tunisia, and USA also exhibited quinolone resistance (Costa et al., 2009; Doi et al., 2010; Jouini et al., 2007; Ojo et al., 2016; Tate et al., 2017). Colistin-resistant *E. coli* isolates harboring β-lactamase and *mcr-1* genes were isolated from chicken meat samples in Brazil, Denmark, the Netherlands and Switzerland (Hasman et al., 2015; Monte et al., 2017; Zogg et al., 2016). Avian pathogenic *E. coli* from the US, China and Egypt were shown to contain *mcr-1* gene (Lima Barbieri et al., 2017).

CATTLE

The occurrence of ESBL/AmpC-*E. coli* in cattle suffering with diarrhea, urinary tract infection, colibacillosis and other diseases varied from 0.4% to 70.0% and the most common β-lactamase group was CTX-M (Table 2) (Asai et al., 2011; Brinas et al., 2005; Kirchner et al., 2011; Liebana et al., 2006; Meunier et al., 2010; Meunier et al., 2006). During 1996-1998, 13.0-69.0% of ceftiofur-resistant *E. coli* isolates were recovered from cows suffering with diarrhea in USA (Bradford et al., 1999; White et al., 2000). The first *Salmonella* strain (*S.* Typhimurium DT104) resistant to cephalosporins was isolated during 1995-1999 from a fecal sample of cattle among 8,476 Canadian animal isolates and possessed CMY-2 and TEM-1 (Allen and Poppe, 2002). Since then many reports of ESBL/AmpC-encoding *E. coli* or *Salmonella* from healthy cattle have been published (Carattoli, 2008). Although US FDA prohibited cephalosporin use in livestock in 2012, ceftriaxone-resistant *Salmonella* isolated from ground beef in USA increased from 11.1% in 2011 to 26.7%-38.5% in 2013-2014 (FDA, 2014). Furthermore, ceftriaxone-resistant *E. coli* isolates obtained from ground beef also increased from 0.5% in 2011 to 0.5-2.2% in 2013-2014. A very low percentage of cefotaxime-resistant *E. coli* were detected from fecal samples

(2.0%, 1/50) of cattle and retail beef meat (1.0%, 1/100) in Japan and harbored CMY-2 (Hiroi et al., 2011). In Tunisia, Jouini et al. screened ESBL-expressing *E. coli* among 16 cattle and 23 beef meat samples and found 5 isolates in the beef (Jouini et al., 2007). Prevalence of ESBL/AmpC-expressing *E. coli* isolates in beef in Tunisia yielded only one ESBL (7.1%, 1/14) and one AmpC producer (7.1%, 1/14) (Ben Slama et al., 2010). Kim and Wei investigated the incidence of *E. cloacae* isolates expressing AmpC gene in cattle farms, beef processing plant and ground beef, and their numbers were the highest in beef processing plant (93.8%, 15/16), followed by ground beef (85.7%, 36/42) and cattle farms (60.0%, 3/5) (Kim and Wei, 2007).

ESBL/AmpC-Associated Genes in Cattle

The detection rate of cephalosporin-resistant *E. coli* isolates varied from 0% to 42.0% and CTX-M-1 has been reported as a major ESBL allele. In North America, CMY-2 has been predominant in most studies. Wittum et al. reported that 6.0% (3/50) of ESBL-encoding *E. coli* strains isolated from feces of bovine samples in USA contained the CTX-M-1 and CTX-M-79 (Table 2) (Wittum et al., 2010). In another report, the prevalence of CTX-M- and CMY-2-encoding *E. coli* was described as 9.4% (70/747) and 95.7% (715/747), respectively (Mollenkopf et al., 2012). In Tunisia, an ESBL producing *E. coli* from beef was reported to carry CTX-M-1, and AmpC producer had mutations at position -42, -18, -1 and $+58$ in the *AmpC* gene promoter. Doi et al. compared the occurrence of ESBL- and CMY-2-encoding *E. coli* in retail meat in Pittsburgh, USA, and Seville, Spain, and found that one ground beef sample (8.3%, 1/12) in Seville was positive for ESBL-producing *E. coli* and one ground beef sample (5.0%, 1/20) in Pittsburgh was positive for CMY-2-producing *E. coli* (Doi et al., 2010). Third-generation cephalosporin-resistant *E. coli* strains were obtained in 2.0% (1/50) of beef in Switzerland (Vogt et al., 2014). The beef isolate produced CTX-M-1 located on IncI1 plasmid. In 2015 European Union survey, 5.0% and 1.8% of ESBL- and AmpC-producing *E. coli* isolates, respectively, were recovered from beef (European Food Safety Authority, 2017).

Zhao et al. reported that 34 and 7 bacteria expressing CMY-4 were detected in US cattle and ground beef samples, respectively (Zhao et al., 2001). The types of β-lactamase genes detected by PCR were CTX-M-1 and TEM-1. In the UK, recovery rate of cefotaxime/ceftazidime-resistant *E. coli* from dairy farm was 42.1% (53/126) (Ibrahim et al., 2016) and the isolates contained the CTX-M, TEM and OXA. In another study, CMY-2-producing *E. coli* was isolated from a fecal sample of cattle and possessed conjugative plasmids that could not be differentiated by restriction fragment length polymorphism analysis (Batchelor et al., 2005). In cattle feces, 4.8% (13/271) of ESBL producers were isolated and possessed CTX-M-1 and TEM-71 (Hartmann et al., 2012). Their repetitive element sequence-based PCR (rep-PCR) profiles showed high diversity between the isolates of different farms. Interestingly, CTX-M was also detected in isolates recovered from soil (18.3%, 22/120). *E. coli* strains carrying CTX-M-1, CTX-M-14, and CTX-M-15 were recovered from eight cattle farms and the serotype of CTX-M-15-encoding *E. coli* was determined as O25 that correlated with urinary tract infection in human (DEFRA, 2013; Molina-Lopez et al., 2011). The frequency of ESBL-expressing *E. coli* from fecal swabs of Swiss cattle was 3.9% (2/51) and carried only two β-lactamase genes, CTX-M-1 and CTX-M-15 (Endimiani et al., 2012c). In Asia, the first CTX-M-producing *E. coli* isolates were isolated from cattle slaughter plants in Japan between 2000-2001 (Shiraki et al., 2004). Six of 396 fecal samples (1.5%) and 2 of 270 carcass swabs have been shown to harbor CTX-M-2. Other Japanese researchers reported that two ceftiofur-resistant *E. coli* isolates (0.2%, 2/1,095) were isolated from feces and harbored a mutant version of the SHV-12 gene with six mutations (Hiki et al., 2013). In Spain, among 459 *E. coli* isolates from sick animals, two bovine clinical *E. coli* strains obtained from milk and fecal samples produced CTX-M-1 and TEM-1b (Brinas et al., 2005). Three clinical bovine *E. coli* isolates possessing ESBL were recovered from fecal and urinary tract samples in France and their ESBL variants were CTX-M-1 and CTX-M-15 that were flanked by the insertion element IS*Ecp1* (Meunier et al., 2006). The same group reported that three ceftiofur-resistant *E. coli* isolated from cattle with respiratory and digestive disease showed resistance to florfenicol and carried IncA/C plasmids disseminating both CMY-2 β-lactamase and *floR* genes (Meunier et al., 2010). In the UK, ESBL-bearing *E. coli* strains were detected in the range of 28.2% to 51.0% from fecal samples of cattle with diarrhea depending on farm visits (Liebana et al., 2006). The CTX-M was found more in calves (64.6%, 31/48) than in cows (3.3%, 2/60). *E. coli* encoding CTX-M-1 group was isolated from liver, lung, spleen and fecal samples of cattle in the UK Veterinary Laboratories Agency (Kirchner et al., 2011). The most commonly found ESBL variant was CTX-M-15/28 and these CTX-M genes were carried on IncI1, IncF and IncA/C plasmids. In Japan, cefazolin-resistant *E. coli* isolated from cattle with colibacillosis was 8.3% (6/72) and carried CTX-M-2 (Asai et al., 2011).

ESBL/AmpC-encoding *Salmonella* have also been found in sick cattle and reported to frequently carry the CMY-2. Between 1998 and 1999, four cephalosporin-resistant *Salmonella* were isolated from lung, lymph node, intestine and fecal samples in sick cattle in Canada and expressed CMY-2 (Winokur et al., 2000). Authors believed that the detection of CMY-2 variant was probably related to the use of ceftiofur to treat respiratory disease. *S.* Typhimurium variant Copenhagen harboring TEM-1 and CMY-2 was found in calves with a severe diarrhea in USA (Fey et al., 2000). In Japan, Sugawara et al. isolated cephalosporin-resistant *S.* Typhimurium from cattle suffering from fever, diarrhea and septicemia in 2007 and it harbored the CMY-2 gene present on 95 kb-IncI1-Iγ and >165 kb-IncA/C plasmids (Sugawara et al., 2011). Only two *Salmonella* strains (*S.* Anatum and *S.* Infantis) producing ESBL recovered from cattle were confirmed among 22,679 isolates of German National Salmonella Reference Laboratory Collection during 2003-2007 and carried CTX-M-1 located on IncI1 and IncN plasmids (Rodriguez et al., 2009). Alcaine et al. found that all ceftiofur-resistant *Salmonella* obtained from US dairy farms had CMY-2 but none of ceftiofur-sensitive *Salmonella* had CMY-2 (Alcaine et al., 2005). Additionally, between 2000 and 2005, the occurrence of *S.* Typhimurium to carry CMY-2 was investigated in humans, retail meat and intestines of food producing animals in Mexico (Zaidi et al., 2007). The AmpC producers were very low (0-3.0%) in retail beef and bovine intestines. CMY-2-expressing *S.* Typhimurium isolates were the most frequently found in children with diarrhea (39.1%), followed by retail pork (19.1%) and swine intestines (17.4%).

Cattle ESBL/AmpC-Producers Other Than *E. Coli* and *Salmonella*

French scientist found that the prevalence of ESBL-expressing Enterobacteriaceae in healthy cattle was higher than in sick animals (Table 2) (Madec et al., 2008). They screened ESBL-expressing Enterobacteriaceae strains (*E. coli, Acinetobacter* spp., *P. aeruginosa, C. freundii, Hafnia alvei*) grown onto ceftazidime- or cefotaxime-supplemented media and recovered 4.1% (25/607) and 2.6% (17/657) from healthy and sick cattle, respectively. These *E. coli* isolates encoding ESBL contained CTX-M-1, CTX-M-14, CTX-M-15, SHV-12 and TEM-126.

Multidrug-Resistance in ESBL/AmpC-Producers in Cattle

TEM-52-expressing *E. coli* isolate was isolated from sliced beef in Denmark and exhibited resistance to several antibiotics including nalidixic acid, ciprofloxacin, sulphonamides, tetracycline and trimethoprim (Table 2) (Jensen et al., 2006). Tate et al. sequenced whole genomes of CTX-M-65-producing *S.* Infantis isolate from US beef (Tate et al., 2017). The beef isolate harbored a single mutation (D87Y) of *gyrA* and many antimicrobial resistance genes, including *aph(4)-Ia, aph(3')-Ic, aac(3)-IVa, aadA1, fosA3, floR, sul1, tetA,* and *dfrA14*. Authors first found CTX-M-65 and pESI-like megaplasmid from *S.* Infantis in USA. *S.* Newport strains resistant to more than 9 antibiotics were isolated from sick food-producing animals, including cattle, chickens, pigs, and turkeys, in USA and produced plasmid-encoded CMY-2 (Zhao et al., 2003).

Milk

Milk, particularly from cows suffering from clinical mastitis, has been considered as a source of contamination of ESBL/AmpC-producing bacteria. Among Enterobacteriaceae, *E. coli* has been reported as a major ESBL/AmpC producer in mastitic milk. A low number of ESBL-expressing *E. coli* isolates were obtained from mastitic milk samples in India and Turkey (Table 2) (Kar et al., 2015). In Tunisia, one ESBL-producing *E. coli* was isolated from milk of cattle with clinical mastitis (Grami et al., 2014). Locatelli et al. isolated 22 Gram-negative bacteria from lactating cows with clinical mastitis in Italy and found only one *E. coli* harboring ESBL (Locatelli et al., 2009). About ten percent (n=82) of Enterobacteriaceae expressing ESBL was isolated from 866 bulk tank milk samples in Germany (Odenthal et al., 2016). ESBL producers included *E. coli* (75.6%), followed by *Citrobacter* spp. (9.6%), *E. cloacae* (6.1%), and *Klebsiella oxytoca* (3.7%). ESBL-encoding bacteria were also detected in waste milk containing antibiotic residues. In waste milk collected from 103 UK farms, cefalexin, cefalonium, cefapirin and cefquinome residues were detected in 3.0-21.0% of the samples and ESBL-encoding bacteria were found in 33.0-74.0% of these samples (Randall et al., 2014a). The prevalence of ESBL-encoding *K. pneumoniae* isolates was studied in healthy cow and mastitic milk in India and more ESBL producers were recovered from mastitic milk (91.3%, 21/23) than healthy cow milk (8.7%, 2/23) (Koovapra et al., 2016). Swiss researchers found 13.7% (17/124) ESBL-encoding bacteria in cattle, 25.3% (16/63) in calves and 1.5% (1/67) in mastitis milk, and no ESBL producers in bulk tank milk and minced meat (Geser et al., 2012).

Table 2. Presence of ESBL/AmpC-producing Enterobacteriaceae in cattle

Species	Sample source	Year	Country	% (Sample #)	CTX-M	TEM	SHV	AmpC and other β-lactamase	Other resistance Phenotype	Other resistance Genotype	MLST	Inc group	Reference
E. coli	Sick	1996	USA	77 (32)		TEM			CHL, CIP, GEN, KAN, NAL, STR, SUL, SXT, TET				(Bradford et al., 1999)
S. Typhimurium	Sick	1998	USA	4 isolates from diarrheal disorder				CMY-2					(Fey et al., 2000)
S. Heidelberg, S. Give, S. Typhimurium, S. Typhimurium subsp. Copenhagen	Sick (cattle & pig)	1998-1999	USA	2.5 (377)		1		CMY-2	CHL, CIP, GEN, NAL, STR, SUL, SXT, TET				(Winokur et al., 2000)
E. coli	Sick	1998-1999	USA	22.2 (189)				CMY-2	CHL, CIP, GEN, SUL, STR, TET, TOB				(Winokur et al., 2001)
S. Typhimurium	Healthy	1994-1999	Canada	0.01 (8426)		1		CMY-2	CHL, FLO, KAN, NEO, STR, SUL, TET				(Allen and Poppe, 2002)
S. Newport	Healthy	2000-2001	USA	34 isolates				CMY-2	CHL, GEN, KAN, SPE, STR, SUL, TET	aadA1			(Rankin et al., 2002)
S. Newport	Sick	1999-2001	USA	0.6 (2252)				AmpC					(Gupta et al., 2003)
E. coli	Healthy/meat	2000-2001	Japan	1.5 (396)-healthy; 0.7 (270)-meat	2								(Shiraki et al., 2004)
E. coli	Healthy	2003	USA	95.9 (122)				CMY-2	CHL, FLO, Gen, SPE, TET				(Donaldson et al., 2006)

Table 2. (Continued)

Species	Sample source	Year	Country	% (Sample #)	CTX-M	TEM	SHV	AmpC and other β-lactamase	Other resistance Phenotype	Other resistance Genotype	MLST	Inc group	Reference
E. coli	Healthy	2003	USA	95.9 (122)				CMY-2	CHL, FLO, Gen, SPE, TET				(Donaldson et al., 2006)
E. coli	Meat	2002	China	3.1 (96)	13	1b			CHL, CIP, GEN, SUL, TET, TMP				(Duan et al., 2006)
E. coli	Healthy	2003-2005	UK	4 ESBL-producing isolates	14			CMY-2	SUL			I1, K	(Hopkins et al., 2006)
E. coli	Meat	2004	Denmark	0.4 (235)		52			CIP, NAL, SUL, TET, TMP				(Jensen et al., 2006)
E. coli	Healthy/ environment	2004-2005	UK	37.0 (297)	CTX-M			AmpC					(Liebana et al., 2006)
E. coli	Sick	2004	France	3 ESBL-producing isolates	1, 15	1			CHL, ENR, FLO, GEN, KAN, NAL, STR, SUL, TET, TMP				(Meunier et al., 2006)
E. coli	Meat	2006	Tunisia	21.7 (23)	1	1	5		CHL, GEN, STR, SUL, TET	aadA1, aac(3)-II, aad B, dfrA1, sul1, tet(A), tet(B)			(Jouini et al., 2007)

Species	Sample source	Year	Country	% (Sample #)	CTX-M	TEM	SHV	AmpC and other β-lactamase	Other resistance Phenotype	Other resistance Genotype	MLST	Inc group	Reference
Enterobacter cloacae	Healthy/meat/ farm & plant environment	2004-2005	USA	43.0 (149)				AmpC					(Kim and Wei, 2007)
S. Typhimurium	Meat	2000-2005	Mexico	0.9 (336)				CMY-2					(Zaidi et al., 2007)
S. Heidelberg	Healthy/meat	2002, 2004	Canada	3 isolates				CMY-2	CHL, GEN, KAN, STR, SUL, TET	aadA1, dhfrA1, floR, strA, sul1, tet(A)		I1	(Andrysiak et al., 2008)
S. Agona, S. Cerro, S. Derby, S. Dublin, S. Heidelberg, S. Infantis, S. Newport, S. Ohio, S. Reading, S. Saint Paul, S. Thompson, S. Typhimurium var 5–, S. Typhimurium, S. Uganda	Healthy	2000-2004	USA	2.4 (3984)		TEM		CMY-2	CHL, GEN, KAN, STR, SUL, SXT, TET				(Frye et al., 2008)
E. coli	Healthy/sick	2005-2006	France	4.1 (607)	1, 14, 15	126	12						(Madec et al., 2008)
E. coli	Milk (sick)	Not specified	Italy	22.7 (22)	1	TEM							(Locatelli et al., 2009)
E. coli	Healthy	Not specified	Canada	2.1 (2483)				CMY-2	CHL, GEN, KAN, STR, SUL, SXT, TET	tet(A), tet(B)		A/C	(Mulvey et al., 2009)

Table 2. (Continued)

Species	Sample source	Year	Country	% (Sample #)	CTX-M	TEM	SHV	AmpC and other β-lactamase	Other resistance Phenotype	Other resistance Genotype	MLST	Inc group	Reference
S. Agona, S. Kinshasa, S. Newport, S. Reading	Healthy/meat	Not specified	USA	146 isolates				CMY-2	CHL, GEN, KAN, STR, SUL, TET			A/C, B/O, HI1, N	(Poole et al., 2009)
S. Anatum, S. Infantis	Healthy/meat	2003-2007	Germany	3 ESBL isolates	1	52							(Rodriguez et al., 2009)
E. coli	Meat	2007	Tunisia	7.1 (14 food samples)	1				CIP, NAL, SUL, SXT, TET	aadA5, dfrA17, sul2, tet(B)			(Ben Slama et al., 2010)
E. coli	Healthy	2010	Denmark	10 (192)	1, 8, 14, 15			AmpC					(DANMAP, 2010)
E. coli	Sick	2006-2007	UK	1.3 (3027)	1, 3, 14, 15, 20, 32								(Hunter et al., 2010)
K. pneumoniae	Milk (sick)	2008-2009	Italy	6.4 (140)	1	TEM	SHV		DAN, KAN, STR, SXT, TET				(Locatelli et al., 2010)
E. coli	Healthy/sick	2009	USA	8 (50)	1, 79				CHL, KAN, NAL, SXT, TET				(Wittum et al., 2010)
E. coli	Sick	2001-2006	Japan	2.8 (72)	2								(Asai et al., 2011)
E. coli, C. youngae	Healthy	2009	Switzerland	17.1 (64)	CTX-M-1, -9 group	TEM			CHL, CIP, GEN, STR, TET				(Geser et al., 2011)
E. coli	Healthy	2004-2006	Japan	2.2 (45)				CMY-2					(Hiroi et al., 2011)
E. coli	Healthy	2008-2010	China	32.9 (210)	14, 28, 55, 98				AMI, CHL, CIP, GEN, NAL, TET, TRI				(Ho et al., 2011)

Species	Sample source	Year	Country	% (Sample #)	CTX-M	TEM	SHV	AmpC and other β-lactamase	Other resistance Phenotype	Other resistance Phenotype (genotype)	MLST	Inc group	Reference
E. coli	Healthy	Not specified	UK	0.013% (median values for the % of E. coli organisms containing blaCTX-M gene)	14, 15								(Horton et al., 2011)
E. coli	Healthy /sick/ milk (sick)	2006-2007	UK	not specified	1, 3, 15/28				CIP, GEN, NAL, SXT, STR, SUL, TET	aac6-1b, aadA4, ant21a, catA1, catB3, dfrA17, floR, oxa1, sul1, sul2, strA, strB, tem, tetA, tetB		A/C, F, FIA, FIB, I1, N	(Kirchner et al., 2011)
S. Typhimurium	Sick	2010	France	not specified	1					aadA5, dfrA17		I1	(Madec et al., 2011)
E. coli	Sick	2007-2008	UK	6.9 (101)	CTX-M-1 group, CTX-M-14								(Snow et al., 2011)
S. Typhimurium	Sick	2007	Japan	2 ESBL-producing S. Typhimurium		1		CMY-2	CHL, GEN, KAN, NAL, STR, TET			A/C, FIB, FII, I1-1γ	(Sugawara et al., 2011)
E. coli	Healthy	2011	Germany	0.7 (2000)	1, 15	52							(Wieler et al., 2011)

Table 2. (Continued)

Species	Sample source	Year	Country	% (Sample #)	CTX-M	TEM	SHV	AmpC and other β-lactamase	Other resistance Phenotype	Other resistance Genotype	MLST	Inc group	Reference
E. coli	Meat	2009	Denmark & other European countries	0.7 (142)-Denmark, 1.2 (84)-Import	1								(Agerso et al., 2012)
E. coli	Milk (sick)	2010	Switzerland					CMY-2	GEN, KAN, NEO, SUL, TET, TOB	aac(3)-IIc, aph(39)-Ic, sul2, tet(A)	66		(Endimiani et al., 2012a)
E. coli, Citrobacter youngae	Healthy/milk	2010-2011	Switzerland	13.7 (124)-healthy, 1.5 (67)-milk	1, 14, 15, 117	1			CHL, CIP, GEN, NAL, STR, SXT, TET				(Geser et al., 2012)
E. coli	Healthy	2007	Japan	12.5 (16)									(Hiroi et al., 2012)
E. coli	Healthy/environment	2009-2010	UK	41.5 (65)	CTX-M-1 group (CTX-M-1, -15, -32, -55), CTX-M-2 group (CTX-M-2), CTX-M-9 group (CTX-M-14, -14B, -27)							N, Y	(Snow et al., 2012)
E. coli	Sick	2006-2010	France	204 ESBL-producing E. coli	CTX-M-1 group (CTX-M-15), CTX-M-2, -9 group	52			CHL, ENR, FLO, GEN, KAN, NAL, STR, SUL, SXT, TET, TRI, TOB				(Valat et al., 2012)

Species	Sample source	Year	Country	% (Sample #)	CTX-M	TEM	SHV	AmpC and other β-lactamase	Other resistance Phenotype	Other resistance Genotype	MLST	Inc group	Reference
E. coli	Healthy/ environment	2008-2009	UK	17.0 (1292)	15								(Watson et al., 2012)
E. coli, K. pneumoniae	Sick	2009-2011	France	0.4 (1427)	1, 14	TEM			APR, CHL, ENR, GEN, KAN, NAL, NET, OFL, STR, SUL, TET, TOB, TRI		23, 45, 10, 58	B/O, I1, F, FIA, FIB, N, Y	(Dahmen et al., 2013)
E. coli	Healthy	2011	Switzerland	One novel CTX-M group 1 variant (CTX-M-117)	117						367		(Hachler et al., 2013)
E. coli	Healthy	2004-2009	Japan	0.2 (1095)			12	AmpC	SXT, TET				(Hiki et al., 2013)
C. koseri, E. aerogenes, E. coli, K. oxytoca, K. pneumoniae	Milk (sick)	2007-2011	Japan	0.22 (28900)	2, 14, 15	1	1, 11, 28, 52, 83, 92, 98, 108, 148	OKP-A	CHL, CIP, ENR, GEN, KAN, LEV, OXY, SXT		10, 23, 88, 101, 155, 540, 648, 1126, 1167, 1284, 1415, 2325, 3499		(Ohnishi et al., 2013)
E. coli, Enterobacter cloacae, Citrobacter youngae	Healthy	2010-2011	Switzerland	8.4 (571)					CHL, CIP, GEN, NAL, STR, SXT, TET				(Reist et al., 2013)

Table 2. (Continued)

Species	Sample source	Year	Country	% (Sample #)	CTX-M	TEM	SHV	AmpC and other β-lactamase	Other resistance Phenotype	Genotype	MLST	Inc group	Reference
E. coli	Healthy /environment	2011-2012	Germany	32.8 (598)	CTX-M-1, -2, -9 group	52		CMY-2, FOX, AmpC	AMI, CIP, GEN, SXT, TOB				(Schmid et al., 2013)
E. coli	Healthy	2008-2009	Korea	0.2 (654)	15							F, FIB	(Tamang et al., 2013b)
E. coli	Healthy/ milk/ environment	2008	Korea	5.5 (1536)	14, 15, 32	1			APR, CHL, CIP, GEN, NEO, NAL, STR, SXT, TET			F, FIB, I1-Iγ, N, P	(Tamang et al., 2013a)
E. coli	Healthy/flies	2010	Japan	10.7 (252)	CTX-M-1 group (CTX-M-15), blaCTX-M-2 group	TEM				tetA	38	FIB	(Usui et al., 2013)
E. coli	Meat	2009-2011	Denmark	3.7 (Not specified)	1	52		CMY-2					(Carmo et al., 2014)
E. coli	Meat	2010-2011	Sweden	0-8.0 (Not specified)	1, 15	1, 52			CHL, CIP, KAN, NAL, GEN, STR, SUL, TET, TRI			F, FIIA, HI2, I1	(Egervarn et al., 2014)
Citrobacter, Enterobacter cloacae, E. coli, Kluyvera intermedia, Raoultella terrigena	Waste milk	2011	UK	6.8 (103)	1, 14, 14b, 15							I1-γ, N	(Randall et al., 2014a)

Species	Sample source	Year	Country	% (Sample #)	CTX-M	TEM	SHV	AmpC and other β-lactamase	Other resistance Phenotype	Other resistance Genotype	MLST	Inc group	Reference
K. pneumoniae	Milk (sick)	2011	Japan	3 ESBL-producing K. pneumoniae	2				KAN, OXY				(Saishu et al., 2014)
E. coli	Meat	2013	Switzerland	2.0 (50)	1				SUL, TET, TRI	dfrA17, IntI1, sul2, tet(B)	453	I1	(Vogt et al., 2014)
E. coli, K. pneumoniae	Healthy	2013	Israel	23.7 (1226)	blaCTX-M-1, -9 group								(Adler et al., 2015)
E. coli	Healthy	2015	Denmark	7.8 (180)	1, 14	52b		AmpC					(DANMAP, 2015)
E. coli	Meat	2015	Denmark	2.5 (315)	1, 14, 15, 32								(DANMAP, 2015)
E. coli	Healthy/sick	2006-2007	Denmark	4 CTX-M-1-positive E. coli	1							I1, N	(Jakobsen et al., 2015)
E. coli	Milk (sick)	2013-2014	India	3.1 (64)	CTX-M	TEM	SHV	AmpC	AMI, GAT, LEV, LOM, NOR, TOB	int1			(Kar et al., 2015)
E. coli	Healthy	2011-2012	Korea	26.9 (290)		1						B/O, FIA, FIB, Frep, HI1, I1, N, P, Y	(Shin et al., 2015)
K. pneumoniae	Milk (healthy)	2011	Indonesia	25.0 (80)	15	1	1, 2a, 11, 28, 36						(Sudarwanto et al., 2015)

Table 2. (Continued)

Species	Sample source	Year	Country	% (Sample #)	CTX-M	TEM	SHV	AmpC and other β-lactamase	Other resistance Phenotype	Other resistance Phenotype	MLST	Inc group	Reference
E. coli	Healthy /environment	2014	Egypt	42.8 (266)	1, 9, 15	TEM	SHV	CMY, OXA-1, -7, -10, -48, -181, MOX-CMY-9	CIP, COT, GEN, MOX, TET, TIG, TOB	aac(3')-IVa, aac(6')-Ib, aadA1, aadA2, aadA4, aadB, ant2, aphA, catA1, catB3, cmlA1, dfrA1, dfrA5, dfrA7, dfrA12, dfrA14, dfrA15, dfrA17, dfrA19, ermB, floR, mphA, mrx, qnrA1, qnrS, qepA, strA, strB, sul1, sul2, sul3, tet(A), tet(B), tet(D)			(Braun et al., 2016)
E. coli	Healthy	2011	Netherlands	41.0 (100)	1, 2, 14, 15, 32, 55	1a, 1b, 52		CMY-2					(Gonggrijp et al., 2016)
E. coli	Healthy/ environment	2010	UK	86.0 (1409)	1, 14, 15								(Horton et al., 2016)
E. coli	Healthy	2012-2014	UK	42.1 (126)	CTX-M	TEM		OXA-1	CHL, CIP, ENR, NAL, OXY, STR, SUL, SXT				(Ibrahim et al., 2016)
K. pneumoniae	Milk (healthy/sick)	Not specified	India	6.7 (340)	15, 63	1	180	AmpC	CHL, CIP, GAT, GEN, NOR, SXT, TET, TOB.	aadA2, aadA5, dfrA12, dfrA17, qnrB, qnrS, sul1			(Koovapra et al., 2016)
A. baumannii, C. braakii, C. freundii, C. youngae, E. coli, E. cloacae, E. amnigenus, H. alvei, K. oxytoca, R. ornitholytica, S. odorifera	Milk (healthy)	2011-2012	Germany	9.5(866)	CTX-M-1, -2, -8, -9 group, CTX-M-1, -2, -15	52	SHV	AmpC, KPC					(Odenthal et al., 2016)

Species	Sample source	Year	Country	% (Sample #)	CTX-M	TEM	SHV	AmpC and other β-lactamase	Other resistance Phenotype	Genotype	MLST	Inc group	Reference
E. coli	Milk (sick)	2015	Turkey	3 ESBL-producing E. coli	15	1			CHL, CIP, ENR, GEN, KAN, NAL, STR, SXT, TET				(Pehlivanoglu et al., 2016)
E. coli	Milk (sick)	2014–2016	Germany	5.5 (490)	1, 2, 14, 15, 32	1			CHL, ENR, GEN, MAR, SXT, TET, TOB	aac(6')-1b-cr	10, 167, 410, 744		(Eisenberger et al., 2018)
E. coli	Milk (sick)	2009–2013	Germany	1.5 (878)	1, 2, 14, 15				CHL, ENR, FLO, GEN, KAN, NAL, STR, SUL, TET, TMP	aac(6')-1b-cr, aadA1, aadA2, dfrA1, dfrA12, floR, sul1, sul2, tet(A)	10, 117, 361, 362, 540, 1431, 1508, 5447	F, FIA, FIB, HI2, I1, N, P	(Freitag et al., 2017)
E. coli	Healthy	2011	Netherlands	13.3 (90)	ESBL			AmpC					(Santman-Berends et al., 2017)
S. Infantis	Healthy	2015	USA	1 blaCTX-M-65-positive S. Infantis	65				CHL, NAL, SUL, SXT, TET	aac(3)-IVa, aadA1, aph(3')-Ic, aph(4)-Ia, dfrA14, floR, fosA3, sul1, tetA			(Tate et al., 2017)

*AMI, amikacin; APR, apramycin; CHL, chloramphenicol; CIP, ciprofloxacin; DOX, doxycycline; ENR, enrofloxacin; FLO, florfenicol; FOS, Fosfomycin; GAT, gatifloxacin; GEN, gentamicin; KAN, kanamycin; LEV, levofloxacin; MIN, minocyclin; NAL, nalidixic acid; NET, netilmicin; NEO, neomycin; OFL, ofloxacin; SMX, RIF, rifampin; sulfamethoxazole; SPE, spectinomycin; STR, streptomycin; SUL, sulphonamides; SXT, sulfamethoxazole/trimethoprim; TET, tetracycline; TMP, trimethoprim; TOB, tobramycin.

**Underline in AmpC means AmpC mutation.

ESBL/AmpC-Associated Genes in Milk

An *E. coli* isolate from mastitic milk sample in Italy was reported to produce CTX-M-1 and TEM (Table 2) (Locatelli et al., 2009). The most frequent allele in ESBL-producing *E. coli* in mastitic milk in Germany was CTX-M-14 (n=10), followed by CTX-M-1 (n=6), CTX-M-15 (n=4), CTX-M-2 (n=1), and CTX-M-32 (n=1) (Eisenberger et al., 2018). In India, the most predominant β-lactamase genes among ESBL-encoding *K. pneumoniae* belonged to CTX-M (82.6%, 19/23), followed by TEM (34.8%, 8/23) and SHV (13%, 3/23) families (Koovapra et al., 2016). In addition, CTX-M gene was located between IS*Ecp1* and *orf477*. In Tunisia, one ESBL-positive *E. coli* from milk of cattle with clinical mastitis was found to harbor the CTX-M-15 located on F2:A-:B- type IncF plasmid (Grami et al., 2014). ESBL-producing *E. coli* from mastitic milk in India and Turkey carried CTX-M, SHV and TEM (Kar et al., 2015).

Multidrug-Resistance in ESBL/AmpC-Producers in Milk

ESBL-producing *E. coli* in mastitic milk samples from Germany were resistant to enrofloxacin and marbofloxacin and some of them contained *aac(6')-Ib-cr*, plasmid-mediated quinolone resistance determinants (Table 2) (Eisenberger et al., 2018). Recently, *E. coli* isolates expressing ESBL were examined from mastitis-positive and –negative milk samples but only 1.4% (12/878) of mastitis-positive samples were found as ESBL producers (Freitag et al., 2017). Apart from possessing several CTX-M genotypes, other plasmid-associated antimicrobial resistance genes, such as *aac(6')-Ib-cr, dfrA1, dfrA12, floR, sul1, sul2*, and *tet(A)* were also found.

ESBL/AmpC-Producers Other Than *E. coli* in Milk

ESBL/AmpC producing bacteria, such as *K. pneumoniae, K. oxytoca, Citrobacter* spp. and *E. cloacae* have been recovered from mastitic milk samples. Ohnishi et al. recovered 419 cefazolin-resistant Enterobacteriaceae (*K. pneumoniae, K. oxytoca, Citrobacter koseri, E. coli, Enterobacter aerogenes*) from about 260,000 milk produced by dairy cows with clinical mastitis in Japan during 2007-2011 (Table 2) (Ohnishi et al., 2013). Of note, ST10, ST23, and ST58 *E. coli* clones, that produce CTX-M-1 and CTX-M-14, from cows with clinical mastitis in France were the same as those in human infection (Dahmen et al., 2013). Three CTX-M-2-expressing *K. pneumoniae* isolates were detected from cows suffering from clinical mastitis in Japan carried IncT plasmid (Saishu et al., 2014). Among sixty-five ESBL producers, CTX-M-2-expressing *K. pneumoniae* were 63.1% (41/65) and CTX-M-15-expressing *E. coli* were 15.4% (10/65). During 2008-2009, one of the 140 *K. pneumoniae* isolates (0.7%) encoding ESBL was recovered from cows suffering from clinical mastitis in Italy was found to harbor CTX-M-1 gene (Locatelli et al., 2010). In France, Dahmen et al. reported a low prevalence of ESBL-producing *K. pneumoniae* (1.2%, 1/85) from cows with clinical mastitis that produced CTX-M-1 and CTX-M-14 genes (Dahmen et al., 2013).

PIG

ESBL/AmpC-encoding Enterobacteriaceae have been increasingly prevalent in pigs (Madec et al., 2017). During 2010-2011, isolation rate of ESBL-producing *E. coli* strains in fecal swab samples of pigs in Swiss slaughter houses was 3.3% (2/60) (Endimiani et al., 2012c). Geser et al. reported high ESBL producers (15.3%, 9/59) from fecal samples (Table 3) (Geser et al., 2012). In Spain, 36.4% (131/360) of *E. coli* isolates from fattening pig farms produced ESBL/AmpC β-lactamases (Blanc et al., 2006). In Germany, the prevalence of ESBL/AmpC-producing *E. coli* increased from 29% to 45% in healthy swine (Von Salviati et al., 2014). In another report from Germany, 88.2% of pig farms were positive for *E. coli* (Dahms et al., 2015). In Denmark and Finland, a very low percentage (0.7-1.5%) of ESBL/AmpC-bearing *E. coli* was isolated from feces (Paivarinta et al., 2016; Wu et al., 2008). During 2015-2016, 4.6% (16/345) of ESBL-encoding *E. coli* isolates were recovered from fecal samples of healthy pigs in Japan (Norizuki et al., 2017). In the European Union (EU) survey, 2.0% and 0.7% of cephalosporin-resistant *E. coli* and *Salmonella*, respectively, were identified from pigs in 2009 (European Food Safety Authority, 2011). In 2015 European Union survey, occurrence of ESBL- and AmpC-producing *E. coli* recovered from pork was 7.0% and 2.3%, respectively (European Food Safety Authority, 2017). German National Antibiotic Resistance Monitoring (GERM-Vet) program recently reported that ESBL-expressing *E. coli* were found most commonly in diseased cattle (11.2%, 324/2,896), followed by pigs (4.8%, 75/1,562) and chickens (0.8%, 20/2,391) (Michael et al., 2017). During 2015-2016, 4.6% (16/345) of ESBL-encoding *E. coli* isolates were recovered from fecal samples of healthy pigs in Japan (Norizuki et al., 2017). ESBL/AmpC-producing gram-negative bacteria have been reported in sick swine (Aarestrup et al., 2006; Brinas et al., 2005; Cartelle et al., 2004; Jakobsen et al., 2015; Meunier et al., 2006; Michael et al., 2017; Schink et al., 2011). Three ESBL-containing *E. coli* strains were recovered from pigs with urinary tract infection and septicemia in France (Meunier et al., 2006). *Salmonella* were detected in only two countries (Germany: 2.0%, Spain: 0.9%) among 11 EU member countries. Lee et al. identified one *S*. Typhimurium isolate amongst 483 fecal samples of pig with diarrhea and the strain produced CMY-2 and harbored IncA/C as well as IncFIB plasmids (Lee et al., 2014).

ESBL/AmpC-Associated Genes in Pigs

In Switzerland, one CTX-M-1 and one CTX-M-3 were detected among 3.3% (2/60) of *E. coli* ESBL producers in pig feces during 2010-2011 (Endimiani et al., 2012c). In addition, the same group reported that three AmpC-expressing *E. coli* from pig nose carried CMY-2 (Endimiani et al., 2012b). Four different β-lactamase genes (CTX-M-1, CTX-14, TEM-1 and TEM-186) were found from high ESBL producers (15.3%, 9/59) from fecal samples (Table 3) (Geser et al., 2012). In Spain, the most predominant β-lactamases reported in 2006 among *E. coli* isolates from fattening pig farms were CTX-M-1 (69.2%) and CTX-M-14 (45.3%) (Blanc et al., 2006). In 2010, however, 72.4% of ESBL-producing *E. coli* encoding CTX-M-1, CTX-M-9, CTX-M-14, and SHV-12 was observed in feces (Escudero et al., 2010). In a report from Germany, 88.2% of pig farms were positive for *E. coli* and 96.3% of them contained CTX-M (Dahms et al., 2015). In Portugal, the most frequently found β-lactamase gene was CTX-M-1 (Goncalves et al., 2010). Recently, Shin et al. reported that 4.36% of ESBL-encoding *E. coli* strains were colonized in pigs with respiratory and digestive diseases (Shin et al., 2017). Three different variants of CTX-M (CTX-M-14, CTX-M-15, CTX-M-165) were found and CTX-M-15 (77.8%, 7/9) was the most common. *E. coli* encoding CTX-M climbed from 2.2% in 2002 to 10.7% in 2007 in fecal samples in swine farms and CTX-M-15/22 was most frequently detected in livestock in China (Tian et al., 2009). In 2013, the most prevalent ESBL and plasmid replicon types in pigs were CTX-M-15 and IncN harboring CTX-M-55 in food-producing animals in China (Lv et al., 2013).

A comparison between ESBL-producing *E. coli* isolated from rectal swabs from healthy pigs and cattle revealed a higher percentage in healthy pigs (12.6%) than in cattle (5.7%) and the detection rates of CTX-M-14 and CTX-M-65 as 34.3% and 28.6%, respectively (Zheng et al., 2012). In *E. coli* isolates from fecal samples of healthy pigs and cattle in Korea during 2008-2009, a higher prevalence of ESBL-producers was observed in pigs (39.9%) than in cattle (0.3%) and CTX-M-14 (14.7%) was the most common allele of CTX-M class and detected only in pigs (Tamang et al., 2013b). Other CTX-M genes discovered were CTX-M-3, CTX-M-27, CTX-M-55, and CTX-M-65 in Korean livestock. ESBL/AmpC-producing gram-negative bacteria have been reported in sick swine (Aarestrup et al., 2006; Brinas et al., 2005; Cartelle et al., 2004; Jakobsen et al., 2015; Meunier et al., 2006; Michael et al., 2017; Schink et al., 2011). Three ESBL-containing *E. coli* strains were recovered from pigs with urinary tract infection and septicemia in France and harbored CTX-M-1 and TEM-1 genes (Meunier et al., 2006). In Spain, six ESBL-carrying *E. coli* were isolated from gut, liver, lung of ill pigs harboring diverse types of ESBL gene combinations (TEM-1b + CTX-M-14, TEM-1c + CTX-M-14, TEM-1b + SHV-1) were detected (Brinas et al., 2005). Two *E. coli* isolates from swine suffering with diarrhea and septicemia in Denmark were resistant to cephalosporins and harbored CTX-M-1 (Aarestrup et al., 2006). Five *E. coli* isolates encoding CTX-M-1 identified from Danish pigs with diarrhea and other disease harbored IncI1 and IncN plasmids and their MLST types were ST1 and ST49 (Jakobsen et al., 2015). In Germany, one ESBL-expressing *E. coli* isolate out of eighty-seven sick pigs was recovered (Schink et al., 2011). The strain was isolated from a pig with mastitis-metritis-agalactia syndrome and harbored CTX-M-1 and IncN plasmid. In addition, Lim et al. identified a single CTX-M-15-producing *E. coli* strain from a fecal sample of diseased pig (Lim et al., 2009).

ESBL-containing *E. coli* were found more in chicken (43.8%, 7/16) than in pork (8.7%, 4/46) and beef (21.2%, 7/33) in Poland (Wasiński et al., 2013). Chicken meat (83.8%) in Denmark possessed ESBL/AmpC-harboring *E. coli* the most, followed by pork (12.5%) and beef (3.7%) (Carmo et al., 2014). In Japan, ESBL-producing bacteria were detected only in chicken meat, but not in pork or beef (Kawamura et al., 2014). Recently, Enterobacteriaceae expressing ESBL were detected in a relatively high percentage (20.6%) from fresh pork meat in Germany (Schill et al., 2017).

Multidrug-Resistance in ESBL/AmpC-Producers in Pigs

ESBL-producing and colistin-resistant *E. coli* isolated from pigs in Vietnam and Estonia, found to have *mcr-1* gene (Brauer et al., 2016; Malhotra-Kumar et al., 2016).

Table 3. Presence of ESBL/AmpC-producing Enterobacteriaceae in pig

Species	Sample source	Year	Country	% (Sample #)	CTX-M	TEM	SHV	AmpC and other β-lactamase	Other resistance Phenotype	Genotype	MLST	Inc group	Reference
S. Give, S. Heidelberg, S. Typhimurium	Sick	1998-1999	USA	2.5 (158)		TEM-1		CMY-2	CHL, GEN, STR, SUL, SXT, TET				(Winokur et al., 2000)
E. coli	Sick	1998-1999	USA	9.0 (188)				CMY-2	CHL, CIP, GEN, SUL, STR, TET, TOB				(Winokur et al., 2001)
S. Newport	Healthy	2002	USA	1 isolate				CMY-2	CHL, GEN, SPE, STR, SUL, TET	aadA1			(Rankin et al., 2002)
E. coli	Meat	2002	Taiwan	2.0 (50 meat samples)				CMY-2	CHL, CIP, GEN, LEV, NAL, SXT, TOB				(Yan et al., 2004)
E. coli	Healthy	2002	Taiwan	23.3 (30 fecal samples)				CMY-2	CHL, CIP, GEN, LEV, NAL, SXT, TOB				(Yan et al., 2004)
E. coli	Sick	2005	Denmark	2 ESBL-positive isolates	1				APR, GEN, NEO, TET				(Aarestrup et al., 2006)
E. coli	Healthy	2003	Spain	29.8 (131)	1		5, 12		CHL, CIP, GEN, KAN, NAL, SUL, SXT, TET, TMP, TOB			FII, N, Q	(Blanc et al., 2006)
E. coli	Meat	2002	China	2 (300)	3, 14, 22	1b			CHL, CIP, GEN, SUL, TET, TMP				(Duan et al., 2006)
E. coli	Sick	2000, 2004	France	3 ESBL-producing isolates	1	1			CHL, KAN, NAL, STR, SUL, TET, TMP				(Meunier et al., 2006)
S. Rissen	Healthy	1999-2004	Spain	0.2 (436)		1b	12		STR, SUL, TET	aadA, sull, tet(A)			(Riano et al., 2006)

Table 3. (Continued)

Species	Sample source	Year	Country	% (Sample #)	CTX-M	TEM	SHV	AmpC and other β-lactamase	Other resistance Phenotype	Genotype	MLST	Inc group	Reference
E. coli	Healthy/sick	2003-2005	China	1.0 (203)	14	1b			GEN, FLO, KAN				(Liu et al., 2007)
S. Typhimurium	Healthy/meat	2000-2005	Mexico	4.6 (907)				CMY-2					(Zaidi et al., 2007)
S. Heidelberg	Healthy/meat	2004	Canada	3 isolates				CMY-2	CHL, STR, SUL, TET	floR, strA, tet(A)		A/C, I1	(Andrysiak et al., 2008)
Citrobacter freundi	Healthy	1998-2004	Portugal	5.7 (35)	1 (7)	52 (80)	12		KAN, NAL, NET, SPE, STR, TET				(Machado et al., 2008)
E. coli, K. pneumoniae, S. Typhimurium, S. Enteritidis	Sick	1999-2006	Korea	4.5 (156)		1, 1b, 1D, 20, 52	1, 2a, 28, 33	CMY-2, DHA-1					(Rayamajhi et al., 2008)
E. coli	Healthy	2005-2006	Denmark	0.7 (137)	1	1a, 1b		AmpC	SPE, STR, SUL, TET, TMP				(Wu et al., 2008)
E. coli	Healthy	2005-2008	Canada	2.5 (120)				CMY-2	CHL, KAN, STR, SUL, SXT, TET	aac(3)IV, aadA, aphA1, strA/strB, sul1, sul2, sul3, tet(A), tet(B), tet(C)			(Kozak et al., 2009)
E. coli	Healthy	2007	Denmark	80 (70)	1				CHL, FLO, SPE, SXT, TET	aadA, catA1, cmlA, floR, sul1, sul2, sul3, tetA		N	(Moodley and Guardabassi, 2009)
S. Typhimurium	Meat	2003-2007	German	3 ESBL-positive isolates	1	1			KAN, NEO, STR, SUL, SXT, TET, TMP	aadA1, aphA1, dfrA1, strA/B, sul1, sul2, tet(A)			(Ben-Ami et al., 2009)

Species	Sample source	Year	Country	% (Sample #)	CTX-M	TEM	SHV	AmpC and other β-lactamase	Other resistance Phenotype	Other resistance Genotype	MLST	Inc group	Reference
E. coli	Healthy	2006	Belgium	2 ESBL isolates	2, 15				GEN, ENR, FLO, NAL, NEO, STR, SUL, TET, TMP			HI2, I1	(Smet et al., 2009)
E. coli	Healthy/sick	2002, 2007	China	2.2 (90)-2002; 10.7 (122)-2007	15, 22	1	2, 11		AMI, CIP, GEN, NOR, OFL, TET				(Tian et al., 2009)
E. coli	Healthy	2003	Spain	29 ESBL-positive strains	1		5, 12		CIP, NAL		10, 1286		(Cortes et al., 2010)
E. coli	Healthy	2010	Denmark	11 (99)	1, 2			CMY-2, AmpC					(DANMAP, 2010)
E. coli	Healthy	2004	Spain	13.1 (per 160 pigs from 80 farms)	1, 9, 14	1b	12	AmpC	APR, CHL, CIP, FLO, GEN, NAL, STR, SUL, TET, TMP				(Escudero et al., 2010)
E. coli	Healthy	2007	Portugal	24.6 (65)	1	1b			STR, SXT, TET	aadA, strA/strB, tet(A)		FIB, FII, N, P, Y	(Goncalves et al., 2010)
E. coli	Healthy	2009	Switzerland	15.2 (59)	CTX-M-1, -9 group	TEM			CHL, CIP, GEN, STR, TET				(Geser et al., 2011)
E. coli	Meat	2004-2006	Japan	4 (25)				CMY-2					(Hiroi et al., 2011)
E. coli	Healthy	2008-2010	China	63.6 (214)	3, 13, 14, 15, 27, 28, 55, 65				AMI, CHL, CIP, GEN, NAL, TET, TRI				(Ho et al., 2011)

Table 3. (Continued)

Species	Sample source	Year	Country	% (Sample #)	CTX-M	TEM	SHV	AmpC and other β-lactamase	Other resistance Phenotype	Genotype	MLST	Inc group	Reference
E. coli	Healthy	Not specified	UK	0.121% (median values for the % of E. coli organisms containing blaCTX-M gene)	1								(Horton et al., 2011)
E. coli	Sick	2004-2006	Germany	3.0 (33 ampicillin resistant strains)	1				SUL, SXT, TET		1153	N	(Schink et al., 2011)
E. coli	Healthy	2009	Denmark	9.3 (782)	1, 2, 14, 15	20	12	AmpC					(Agerso et al., 2012)
E. coli	Meat	2009	Denmark	2.0 (153) [Denmark], 1.2 (173) [Import]	1, 2	52							(Agerso et al., 2012)
E. coli	Healthy	2010-2011	Switzerland	15.3 (59)	1, 14	1, 186			CHL, CIP, GEN, NAL, STR, SXT, TET				(Geser et al., 2012)
E. coli	Healthy	2007	Japan	3.0 (33)									(Hiroi et al., 2012)

Species	Sample source	Year	Country	% (Sample #)	CTX-M	TEM	SHV	AmpC and other β-lactamase	Other resistance Phenotype	Other resistance Genotype	MLST	Inc group	Reference
E. coli	Meat	2011	USA	1.0 (104)				CMY-2	TET				(Park et al., 2012)
E. coli	Healthy	2002-2007	China	6.6 (212)	15, 22		2, 12	CMY-2	AMI, CIP, GEN, TET				(Tian et al., 2012)
E. coli	Healthy	2004-2009	Japan	0.1 (802)				CMY-2	CHL, GEN, KAN				(Hiki et al., 2013)
E. coli	Healthy	2008-2009	Portugal	49.3 (71)	1, 9, 14, 32	1	12		CHL, CIP, GEN, NAL, STR, SXT, TET, TOB	aac(3)-II, aac(3)-IV, aadA, cmlA, floR, strA, strB, sul1, sul2, sul3, tet(A), tet(B), gyrA, parC	10, 23, 48, 56, 58, 101, 218, 354, 398, 540, 877, 1397, 1508, 1790, 1832, 2524, 2525, 2528		(Ramos et al., 2013)
E. coli	Healthy, farm	2006-2007	Portugal	51.1 (43)	1, 32	1, 52		OXA-1	AMI, CHL, GEN, KAN, NAL, NET, STR, SUL, TET, TOB, TRI		10, 34, 227, 1451	FIA, FIB, FII, I1, N, P, Y	(Rodrigues et al, 2013)
E. coli	Healthy	2008–2009	Korea	21.5 (558)	3, 14, 15, 27, 55, 65	1			CHL, GEN, NEO, STR, SXT, TET			F, FIB, HI1, I1-Iγ, P, X	(Tamang et al., 2013b)
E. coli	Meat	2009-2011	Denmark	12.5 (Not specified)	1, 2, 14			CMY-2					(Carmo et al., 2014)

Table 3. (Continued)

Species	Sample source	Year	Country	% (Sample #)	CTX-M	TEM	SHV	AmpC and other β-lactamase	Other resistance Phenotype	Other resistance Genotype	MLST	Inc group	Reference
S. Derby, S. Rissen, S. Typhimurium	Healthy	2008-2011	Belgium	1.9 (368)	1	52		CMY-2	CHL, STR, SXT, TET				(de Jong et al., 2014)
E. coli	Meat	2010-2011	Sweden	2.0-13.0 (Not specified)	1, 14	1, 52		CMY-2	CHL, CIP, FLO, KAN, NAL, STR, SUL, TET, TRI			A/C, F, I1, N	(Egervärn et al., 2014)
E. coli	Healthy	2010-2011	Denmark	18.6 (339)	1, 14		12				10, 58, 86, 189, 206, 453, 542, 641, 744, 910, 1406, 1684, 2739, 3322, 4048, 4051, 4052, 4053, 4056	F, I1, N	(Hammerum et al., 2014)
S. Typhimurium	Sick	2011-2012	Korea	0.4 (483)				CMY-2	CHL, FLO, GEN, NAL, STR, SXT, TET			A/C, FIB	(Lee et al., 2014)
E. coli	Healthy	2013	UK	23.4 (637)	CTX-M		12						(Randall et al., 2014b)
E. coli	Healthy	2011-2012	Germany	37.0 (420)	CTX-M	1	12	CMY-2					(Von Salviati et al., 2014)
E. coli	Healthy	Not specified	Thailand	44.3 (122)	CTX-M-1, -9 group	1, 135, 176			CHL, CIP, GEN, NAL, NOR, STR, SXT, TET	aadA1, aadA2, aadA22, dfrA5, dfrA12, linF			(Changkaew et al., 2015)

Species	Sample source	Year	Country	% (Sample #)	CTX-M	TEM	SHV	AmpC and other β-lactamase	Other resistance Phenotype	Genotype	MLST	Inc group	Reference
E. coli	Healthy	2015	Denmark	28.6 (273)	1, 14, 55			CMY-2, AmpC					(DANMAP, 2015)
E. coli	Meat	2015	Denmark	1.7 (289)	1, 15			AmpC					(DANMAP, 2015)
E. coli	Healthy/environment	2013	China	120 ESBL-producing E. coli	14, 15, 24, 27, 65	1			CHL, CIP, GEN, KAN, NAL, STR, SXT, TET				(Gao et al., 2015b)
E. coli	Pig farm environment	2014	China	23.3 (180)	13, 14, 15, 27, 65,	1			AMI, CHL, CIP, FLO, GEN, KAN, NAL, TET			F, FIB, K, N, Y	(Gao et al., 2015a)
E. coli	Healthy/sick	2006-2007	Denmark	21 CTX-M-1-positive E. coli	1							I1, N	(Jakobsen et al, 2015)
E. coli	Healthy	2010-2012	China	4.5 (424)	14	1		OXA-1		aac-(6')-Ib-cr, aadA5, drfA17, floR, oqxA, qnrS1, rmtB	10, 101, 156, 359, 457, 648, 1114, 3014, 3244, 3269, 3376, 3402, 3403, 3404		(Guo et al., 2014)
E. coli	Pig farm and receiving river	2014	China	19.0 (300)	14, 15,	52			CIP, ENR, ERY, GEN, KAN, STR, TET				(Li et al., 2015)
E. coli	Healthy	2012	India	6.0 (200)	CTX-M	TEM	SHV		AMI, ENR, GEN, TET			A/C, FIB, K	(Samanta et al., 2015b)
Enterobacteriaceae	Healthy	2012	Germany	30.2 (540)	CTX-M	TEM	SHV	AmpC					(Schmithausen et al., 2015)

Table 3. (Continued)

Species	Sample source	Year	Country	% (Sample #)	CTX-M	TEM	SHV	AmpC and other β-lactamase	Other resistance Phenotype	Other resistance Genotype	MLST	Inc group	Reference
E. coli	Healthy	2013	Taiwan	22 ESBL-producing isolates	CTX-M-1, -9 group	TEM		CMY-2	AMK, CHL, FOS, LEV, TET	fosA3	744, 2310	B/O, FIB, FrepB, I1, N, P	(Tseng et al., 2015)
E. coli	Healthy/ outside & inside the pig barn	2011-2012	Germany	23.1 (247)	1, 9, 15	1	12	CMY-2					(von Salviati et al., 2015)
E. coli	Healthy	2013-2014	China	2 cfr-positive E. coli	14b				AMI, CHL, ENO, FLO, GEN, SUL, TET	cfr	746		(Zhang et al., 2015)
E. coli	Healthy	2011-2012	China	10.8 (93) colistin-resistant isolates	55, 65			mcr-1	CHL, CIP, COL, KAN, NAL, NOR, STR, SXT, TET	aadA1, aadA2, aph(3')-Ia, cmlA1, dfrA12, dfrA14, ermB, floR, mphA, qnrS1, oqxA, strAB, sul1, sul2, sul3, tet(A)		A/C, FIB, FIC, FII, HI2, HI2A, I2, X1	(Bai et al., 2016)
E. coli	Pig Slurry	2011-2014	Estonia		1	1A		mcr-1, AmpC	SXT			X4	(Brauer et al., 2016)
E. coli, E. cloacae	Healthy	Not specified	Canada	138 blaCMY-2-positive E. coli				CMY-2	CHL, GEN, KAN, NAL, STR, SXT, TET			A/C, I1	(Jahanbakhsh et al., 2016)
E. coli	Healthy	2001-2011	Korea	20.3 (469)	CTX-M	TEM	SHV	ACC, CARB, CMY-2, IMP, KPC, NDM-1, OXA, VIM				A/C, B/O, FIA, FIB, Frep, HI1, HI2, I1, N, P, W, Y	(Han et al., 2016)

Species	Sample source	Year	Country	% (Sample #)	CTX-M	TEM	SHV	AmpC and other β-lactamase	Other resistance Phenotype	Other resistance Genotype	MLST	Inc group	Reference
E. coli	Healthy	2015	China	56.7 (60)	14, 15, 55, 65	1			CHL, CIP, GEN, NAL, NEO, STR, SXT, TET		10, 58, 86, 93, 206, 453, 744, 910, 1684, 2936, 4048, 4056		(Zhang et al., 2016)
E. coli	Healthy	2011	Netherlands	19.2 (1350)	1, 2, 14, 15, 32	52							(Dohmen et al., 2017a)
E. coli	Healthy /dust	2011	Netherlands	12.6 (1050)	CTX-M-1 group								(Dohmen et al., 2017b)
E. coli	Healthy /dust	2014	Germany	105 E. coli isolates	1	52			CIP, GEN, SXT		10, 29, 48, 58, 88, 101, 641, 1079, 2496, 2509, 2944, 5122		(Fischer et al., 2017)
Entero-bacteriaceae	Healthy	2015	Switzerland	3.8 (106)	1, 15, 32, 162-like, CTX-M-9 group						10, 58, 59, 131, 205		(Kraemer et al., 2017)
E. coli	Sick	2009-2015	Korea	4.4 (206)	14, 15, 65	1	CMY-2, DHA-1		CIP, GEN, NAL, SXT, TET			FIB, Frep, I1, N	(Shin et al., 2017)
E. coli	Meat	2014-2015	Belgium	47-97%	CTX-M-1, -2, -9 group	TEM	SHV						(Van Damme et al., 2017)
E. coli	Meat	2009-2012	Korea	0.3 (1771)	1, 2, 14/18, 58, 79	1		AmpC	CHL, CIP, GEN, NAL, KAN, STR, TET	aadA1, aadA5, dfrA17, qacEΔ1	88, 101, 156, 167, 180, 2309	B/O, FIA, FIB, HI1, I1-Iγ, P, Y	(Kim et al., 2018)

*AMI, amikacin; APR, apramycin; CHL, chloramphenicol; CIP, ciprofloxacin; DOX, doxycycline; ENR, enrofloxacin; FLO, florfenicol; FOS, Fosfomycin; GAT, gatifloxacin; GEN, gentamicin; KAN, kanamycin; LEV, levofloxacin; MIN, minocyclin; NAL, nalidixic acid; NET, netilmicin; NEO, neomycin; OFL, ofloxacin; SMX, RIF, rifampin; sulfamethoxazole; SPE, spectinomycin; STR, streptomycin; SUL, sulphonamides; SXT, sulfamethoxazole/trimethoprim; TET, tetracycline; TMP, trimethoprim; TOB, tobramycin.

***Underline in AmpC means AmpC mutation.

***Underline in gyrA, gyrB, parC, pare means fluoroquinolone mutation.

ESBL/AmpC-Producers Other Than *E. Coli* and *Salmonella* in Pigs

Rare Enterobacteriaceae species producing ESBL, such as *S.* Rissen, 4,5,12:i:− and Bovismorbificans, *C. freundii,* and *K. pneumoniae,* were reported in Portugal, Spain, and UK (Table 3) (Freire Martin et al., 2014; Machado et al., 2008; Riano et al., 2006). In Korea, ESBL/AmpC-expressing *K. pneumoniae* isolates from sick pigs were found (Rayamajhi et al., 2008). Additionally, ESBL/AmpC-positive Enterobacteriaceae isolated from diseased swine included *E. coli, K. pneumoniae* and *S.* Typhimurium. SHV- and TEM-expressing *E. coli* and *K. pneumoniae* isolates from sick pigs were found in Korea (Rayamajhi et al., 2008).

Reservoirs of ESBL/AmpC–Producing Bacterial Strains and Genes in Food Animals and Their Relevance to Human Health

Poultry meat has been considered a potential reservoir of ESBL-producing Gram-negative bacteria. In Netherlands, antibiotic use in the poultry industry is higher than other European countries (Huijbers et al., 2013). Thus, ESBL/AmpC producers were highest in broilers and poultry meat. Nineteen percent of human isolates harbored poultry-associated ESBL genes that carried by IncI1 (Leverstein-van Hall et al., 2011). In another study, ESBL genes were detected in *E. coli* from eight humans on six pig farms (Dohmen et al., 2015). Human and pig isolates within the same farm harbored similar ESBL gene types and had identical sequence and plasmid types on two farms, suggesting clonal transmission. Human ESBL carriage was associated with average number of hours working on the farm per week and presence of ESBLs in pigs. Transmission of CTX-M-1 harboring IncI1 plasmids between pigs and farmers was also reported in a molecular study on two pig farms (Moodley and Guardabassi, 2009).

An increase in the rate of nosocomial infections due to transmission of ESBL genes from one person to the other in hospital, community and household settings, makes the treatment options limited and the ESBL-producing bacteria as reservoirs of these genes (Control, 2011; Pitout and Laupland, 2008). *E. coli* and *Salmonella* are still the main producers of ESBL/AmpC enzymes and have been predominantly present in food-producing animals in many countries and in foods of animal origin (Bergenholtz et al., 2009; Carattoli, 2008). Their role in the spread of ESBL/AmpC genes has come under increased scrutiny. In 2009, reported prevalence of *Salmonella* in different animal species and meat was found in the range of 0.4% to 5.0% and for *E. coli* as 2.0% to 9.0% in EU countries (European Food Safety Authority, 2011). However, the detection rates of ESBL or AmpC genes in *E. coli* from food-producing animals or food (poultry, swine, bovines, horses, rabbits, ostriches, wild boar) varied from 0.2% to 40% in Portugal, the Netherlands, and France, with slightly lower percentages in other countries (European Food Safety Authority, 2011). Almost 100% of broiler farms and >80.0% of animals tested positive for ESBL *E. coli* in Netherlands (Dierikx et al., 2010b). *Salmonella* isolates exhibited lower prevalence (<1.0%) of ESBLs than for *E. coli* in 2 studies in Germany and Spain, respectively (Lopez-Cerero et al., 2011; Rodriguez et al., 2009). Among the CTX family of ESBL genes, CTX-M-1 is regarded as the most prevalent ESBL gene in food-producing animals (European Food Safety Authority, 2011). However, CTX-M-15, usually present in ESBL-producers in humans, has started showing up in food-producing animals or food (Lopez-Cerero et al., 2011; Mora et al., 2010; Platell et al., 2011). Other classes of ESBL genes, namely SHV-2, SHV-12 and TEM-52 have been frequently seen throughout the European Union (European Food Safety Authority, 2011). AmpC class of β-lactamase was almost always the CMY-2 (European Food Safety Authority, 2011).

ESBL/AmpC genes in both humans and animal ESBL-producers have been found mainly on plasmids IncF, IncI, IncN, IncA/C, IncL/M, that are responsible for their efficient spread (Carattoli, 2009). Plasmids IncN, IncI, and IncL/M have been associated with the spread of CTX-M-1, CTX-M-3, and TEM-52 in Europe (Cloeckaert et al., 2007). IncK plasmids that carry the CTX-M-14 gene have been observed in both Spain and the UK (Cottell et al., 2011). CMY-2 spreads via IncA/C plasmids in both the USA and UK (Hopkins et al., 2006). As the evidence of the existence of common clones of ESBL/AmpC-producing *E. coli* in foods of animal origin and in humans mounted, so were the concerns of zoonotic transfer of *E. coli* O25 clones such as ST131, ST648, ST69 (Cortes et al., 2010; Mora et al., 2011; Platell et al., 2011; Vincent et al., 2010). Other clones of *E. coli* that were widely detected among poultry isolates in EU countries and in Japan include ST57, ST156, and ST371. The most concerning of these ESBL-producing *E. coli* clones included outbreak strains of Shiga toxin–producing *E. coli* clones O111:H8 and O104:H4 from food-producing animals (Bonnedahl et al., 2010; King et al., 2012; Randall et al., 2011; Simoes et al., 2010; Valat et al., 2012). People working with poultry have been demonstrated to have a higher risk for intestinal carriage of ESBL/AmpC-producing bacteria (Dierikx et al., 2010a). Transmission of ESBL/AmpC-producing bacteria from animals to humans is proposed to have occurred via the food chain, either via direct contact with animals, or indirectly via the environment (European Food Safety Authority, 2011; Moodley and Guardabassi, 2009).

Imported chicken meat might be a reservoir for ESBL/AmpC-producing bacteria. Occurrence of *E. coli* isolates encoding ESBL/AmpC in the UK were confirmed from imported chicken meat between 13.4-29.5% (Dhanji et al., 2010; Warren et al., 2008). *E. coli* obtained from chicken meat imported from Argentina, Brazil, Chile, France, Poland and the Netherlands carried CTX-M-2 and CTX-M-8. In Denmark, 1.3% of the 1,650 *E. coli* isolates from imported poultry meat, were ESBL/AmpC producers (Bergenholtz et al., 2009). All ESBL/AmpC producers were imported from Germany, France and Brazil. In Sweden, *E. coli* isolates producing ESBL/AmpC were found in 44 of the 100 chicken meat samples imported from South America (Borjesson et al., 2013). In Ghana and Japan, the observed isolation rates of ESBL/AmpC-containing *E. coli* were higher in imported than in domestic chicken meat (Kawamura et al., 2014; Rasmussen et al., 2015). In Denmark, 36.0%-83.8% *E. coli* bearing ESBL/AmpC were detected predominantly in chicken meat than pork (1.2%-12.57%) and beef (1.2%-3.7%) (Agerso et al., 2012; Carmo et al., 2014). Most chicken meat samples were imported from Germany or France. In Vietnam, ESBL/AmpC-expressing *E. coli* was highest in the chicken meat (92.7%), followed by pork (34.8%), and beef (34.3%) (Nguyen do et al., 2016). While

ESBL/AmpC producers were recovered in a range of 0-8.0% and 2.0-13.0% in pork and beef in Sweden, their prevalence range in chicken meat was from 15.0 to 95.0% (Egervarn et al., 2014). In Japan, 50.0% of chicken meat tested was contaminated with ESBL-producing *E. coli* but none of the pork and beef samples were contaminated with them (Kawamura et al., 2014).

Risk Factors Associated with Emergence and Spread of ESBL/AmpC-Producing Bacterial Strains in Food-Producing Animals and Food

Prior antibiotic use is one of the many risk factors associated with the spread of ESBL/AmpC ß-lactamase-producing bacteria, especially with the oxyimino-ß-lactams (cefuroxime, cefotaxime, ceftriaxone, ceftazidime, or aztreonam), fluoroquinolones and ß-lactam–ß-lactamase inhibitor combinations (Ben-Ami et al., 2009; Harris et al., 2007; Park et al., 2009; Wener et al., 2010). Farm management factors such as animal exposure to contaminated water or feed and absence of water acidification in poultry production may also facilitate the introduction and spread of ESBL/AmpC-producing bacteria (Persoons et al., 2011; Smet et al., 2008). Most ESBL/AmpC genes are carried on transposable genetic elements and conjugative plasmids which also contain resistance genes for multitude of other antibiotics. Therefore, the use of other antimicrobials may also result in the selection and dissemination of ESBL/AmpC-producing bacteria (Persoons et al., 2011; Smet et al., 2008). A strong evidence in this regard comes from a Canadian study where a reduction in ceftiofur-resistant *S. Heidelberg* and *E. coli* from both human infections and retail poultry was observed upon withdrawal of ceftiofur for disease prophylaxis in hatcheries (Dutil et al., 2010). When ceftiofur was partially reintroduced, an increase of resistance in poultry and humans was observed, suggesting possible zoonotic source of ESBL genes.

Direct contact with broilers and pigs has also been identified as an important risk factor for carriage of ESBL-producing Enterobacteriaceae (Dierikx et al., 2013a; Moodley and Guardabassi, 2009). In one study involving 26 farms and 18 farmers in Netherlands, ESBL genes were detected in broiler isolates from 6 farmers on all farms studied. ESBL carriage among farmers was higher (33%) as compared to hospital patients (approximately 5%) (Huijbers et al., 2013; Tangden et al., 2010). Moreover, ESBL isolates from the farmers and broilers from two of these farms had genetic similarities in genes and plasmids (EMA, 2014). In another report, frequent contact with pigs carrying ESBL-producers resulted in ESBL carriage by people working/living on pig farms. ESBL carriage was noted at 13% and 27% among people living/working on a farm and people with daily exposure to ESBL producer-carrying pigs, respectively (Moodley and Guardabassi, 2009). Similarity between ESBL gene types detected in humans and pigs on the respective farms suggested a transmission of ESBL genes from pigs to humans (or vice versa), predominantly via horizontal dissemination of plasmids (Moodley and Guardabassi, 2009; Swedres-Svarm, 2014). However, local recirculation of ESBL/AmpC-producing *E. coli* in poultry production chain and imported breeding chickens have also been documented for vertical transmission of ESBL/AmpC genes (Dierikx et al., 2010a; Mevius et al., 2011; SVARM, 2011). A recent study has clearly shown the transfer of a TEM-52-carrying plasmid from an avian *E. coli* strain to 2 human *E. coli* strains in a continuous flow culture model (Smet et al., 2011). Contamination of chicken meat/meat products with ESBL/AmpC bacteria has also been shown to spread of ESBL/AmpC genes in human population (Doi et al., 2010; Lavilla et al., 2008; Mesa et al., 2006). A high degree of similarity of resistance genes and MLST patterns among ESBL/AmpC bacteria derived from meat and hospitalized patients further confirms this observation (Overdevest et al., 2011). Among a variety of risk factors, hospitalization is also considered as a risk factor for human carriage of ESBLs (Ben-Ami et al., 2009). In a study from USA, it was found that the carriage rate for ESBLs during 2000-2005 increased from 1.3% to 3.2% among high-risk, hospitalized patients and bacteremia developed in 8.5% of all previously identified ESBL gene carriers (Reddy et al., 2007). Similar findings were reported from Spain, where fecal carriage rate in hospitalized patients increased from 0.3% in 1991 to 11.8% in 2003 (Valverde et al., 2004). International travel, especially to India and Asia, has been described as a major risk factor for ESBL colonization (Borjesson et al., 2013; Laupland et al., 2008). A colonization rate of 3.5% was observed among the attendees of an infection control symposium, which was comparable with data from Sweden with 3.0% ESBL colonization in relatively healthy volunteers (Meyer et al., 2012; Stromdahl et al., 2011).

Strategies to Reduce the Public Health Risk Caused by ESBL/AmpC-Producing Bacterial Strains Transmitted via the Food Chain

To develop effective mitigation strategies to reduce the spread of ESBL/AmpC bacteria and risk to human health it is important to identify the risk factors that contribute to the problem and analyze quantitative data on human exposure through the various transmission routes. Some of the proposed recommendations have shown limited success in controlling the spread of ESBLs but since the factors that contribute to their spread are varied in nature, it is impossible that single control method will reduce the spread of ESBL/AmpC-producing bacteria and the associated genes. Public health and safety, therefore, remains a serious concern and effective mitigation strategies at pre-and postharvest levels need to be developed and implemented to minimize the spread of ESBL/AmpC-gens through farm animals and their meat products. Risks to the public health by ESBL/AmpC bacteria are mainly determined by (i) their prevalence in food-producing animals and food, (ii) the genetic make-up of the β-lactamase genes involved, and (iii) the transmission from animals or food to humans and vice-versa. Mitigation strategies should therefore focus on reducing the prevalence of ESBL/AmpC bacteria in animals and food, and to reduce transmission from contaminated animals and foods to humans. The use of cephalosporins and matching antimicrobial compounds has exerted enormous pressure on microorganisms to develop resistance against these compounds. The recommendations proposed here are based on the contributing factors responsible for the spread of ESBLs and how they could possibly be used to develop control measures that might work. Strict measures are required to control the use of these compounds in food-producing animals. Because ESBL producers have been reported most commonly in the poultry industry, control strategies should be focused on chicken production. Measures to control dissemination of ESBL/AmpC genes are also important. The progress should be monitored on a regular basis to determine the effectiveness of any control measures. Restriction of using cephalosporins may decrease the prevalence of ESBL/AmpC-carriers. However, a very restrictive policy might have unintended consequences on animal health and welfare if effective antimicrobials are not available for treatment. Other measures such as minimizing off-label use of cephalosporins and/or decreasing the total antimicrobial use in animal production should also be made a high priority. To illustrate the point, a reduction of 58% in the sales of veterinary antibiotics in Netherlands between 2009 and 2014 resulted in a vastly decreased isolation rates (67%) of *E. coli* from poultry meat in 2014 compared to 2013 (83%) and 2012 (73%), probably due to a reduction in the use of antibiotics (Dierikx et al., 2013a).

Because of the high prevalence of ESBL producers in the chicken production (Dierikx et al., 2010a; Dierikx et al., 2010b; Mevius et al., 2011; SVARM, 2011), antibiotic pressure should be decreased to control the development and dissemination of antimicrobial resistance. One way of minimizing the ESBL producers in food animals is to decrease the number of antimicrobials used in veterinary medicine. To control the increase in ESBL/AmpC in the human healthcare system, precautionary efforts should be made to reduce this prevalence internationally. With a multitude of factors affecting the epidemiology of ESBL/pAmpC genes, a multifaceted approach is required to reduce the spread and transmission of these organisms and associated genes in farm animals and food products derived from them. This would require a regular monitoring of the farm animals, farm workers, food industry, animal waste, run off water, air, hospital patients infected with ESBL/pAmpC-producers and check for any links between human and animal isolates. Precaution should be taken to identify the sources responsible for the spread of ESBL/pAmpC genes via horizontal and vertical transfers. A close co-operation and collaboration among different stakeholders, e.g., farm owners, farmers, food industries, healthcare facilities, environmentalists are essential for the success in reducing the spread and transmission of ESBL/pAmpC genes in humans to maintain the efficacy of cephalosporins.

CONCLUSION

Use of cephalosporins to treat diseased animals on farms has led to increasing prevalence of ESBL/AmpC-producing *E. coli* or *Salmonella* strains in poultry, cattle, and pigs. The presence of similar and sometimes indistinguishable isolates from humans and farm animals suggests a zoonotic transfer and a matter of serious concern for public health. The use of whole genome sequences of ESBL-producing *E. coli* isolates from animals and humans, reconstructed plasmid sequences from the two sources exhibiting indistinguishable features, presence of CTX-M-15 gene (a gene that is usually present in ESBL producers in human ESBL-producing isolates) in animals further strengthen the idea of a possible transmission between both hosts. The presence of plasmid-associated ESBL/AmpC genes along with the virulence and multiple antimicrobial resistance genes observed in some of these isolates raises concerns about their impact on human health and effectiveness of antimicrobial therapy. More cautious use of antibiotics, such as the extended-spectrum cephalosporins, better hygiene and *biosecurity measures in farms and/or improved monitoring and surveillance activities of antimicrobial resistance may help in developing mitigation strategies to limit the prevalence and spread of ESBL/AmpC producers in farm animals and related products.*

DISCLAIMER

The views expressed herein do not necessarily reflect those of the US Food and Drug Administration or the US Department of Health and Human Services.

REFERENCES

Aarestrup, F. M., Hasman, H., Agerso, Y., Jensen, L. B., Harksen, S. and Svensmark, B. 2006. First description of blaCTX-M-1-carrying Escherichia coli isolates in Danish primary food production. *J Antimicrob Chemother.* 57, 1258-1259.

Abdallah, H. M., Reuland, E. A., Wintermans, B. B., Al Naiemi, N., Koek, A., Abdelwahab, A. M., Ammar, A. M., Mohamed, A. A. and Vandenbroucke-Grauls, C. M. 2015. Extended-Spectrum beta-Lactamases and/or Carbapenemases-Producing Enterobacteriaceae Isolated from Retail Chicken Meat in Zagazig, Egypt. *PLoS One.* 10, e0136052.

Abgottspon, H., Stephan, R., Bagutti, C., Brodmann, P., Hachler, H. and Zurfluh, K. 2014. Characteristics of extended-spectrum cephalosporin-resistant Escherichia coli isolated from Swiss and imported poultry meat. *J Food Prot.* 77, 112-115.

Abreu, R., Castro, B., Espigares, E., Rodriguez-Alvarez, C., Lecuona, M., Moreno, E., Espigares, M. and Arias, A. 2014. Prevalence of CTX-M-Type extended-spectrum beta-lactamases in Escherichia coli strains isolated in poultry farms. *Foodborne Pathog Dis.* 11, 868-873.

Adler, A., Sturlesi, N., Fallach, N., Zilberman-Barzilai, D., Hussein, O., Blum, S. E., Klement, E., Schwaber, M. J. and Carmeli, Y. 2015. Prevalence, Risk Factors, and Transmission Dynamics of Extended-Spectrum-beta-Lactamase-Producing Enterobacteriaceae: a National Survey of Cattle Farms in Israel in 2013. *J Clin Microbiol.* 53, 3515-3521.

Agerso, Y., Aarestrup, F. M., Pedersen, K., Seyfarth, A. M., Struve, T. and Hasman, H. 2012. Prevalence of extended-spectrum cephalosporinase (ESC)-producing Escherichia coli in Danish slaughter pigs and retail meat identified by selective enrichment and association with cephalosporin usage. *J Antimicrob Chemother.* 67, 582-588.

Agerso, Y., Jensen, J. D., Hasman, H. and Pedersen, K. 2014. Spread of extended spectrum cephalosporinase-producing Escherichia coli clones and plasmids from parent animals to broilers and to broiler meat in a production without use of cephalosporins. *Foodborne Pathog Dis.* 11, 740-746.

Alcaine, S. D., Sukhnanand, S. S., Warnick, L. D., Su, W. L., Mcgann, P., Mcdonough, P. and Wiedmann, M. 2005. Ceftiofur-resistant Salmonella strains isolated from dairy farms represent multiple widely distributed subtypes that evolved by independent horizontal gene transfer. *Antimicrob Agents Chemother.* 49, 4061-4067.

Aliyu, A. B., Saleha, A. A., Jalila, A. and Zunita, Z. 2016. Risk factors and spatial distribution of extended spectrum beta-lactamase-producing-Escherichia coli at retail poultry meat markets in Malaysia: a cross-sectional study. *BMC Public Health.* 16, 699.

Allen, K. J. and Poppe, C. 2002. Occurrence and characterization of resistance to extended-spectrum cephalosporins mediated by beta-lactamase CMY-2 in Salmonella isolated from food-producing animals in Canada. *Can J Vet Res.* 66, 137-144.

Andrysiak, A. K., Olson, A. B., Tracz, D. M., Dore, K., Irwin, R., Ng, L. K., Gilmour, M. W. and Canadian Integrated Program for Antimicrobial Resistance Surveillance, C. 2008. Genetic characterization of clinical and agri-food isolates of multi drug resistant Salmonella enterica serovar Heidelberg from Canada. *BMC Microbiol.* 8, 89.

Asai, T., Masani, K., Sato, C., Hiki, M., Usui, M., Baba, K., Ozawa, M., Harada, K., Aoki, H. and Sawada, T. 2011. Phylogenetic groups and cephalosporin resistance genes of Escherichia coli from diseased food-producing animals in Japan. *Acta Vet Scand.* 53, 52.

Bai, L., Hurley, D., Li, J., Meng, Q., Wang, J., Fanning, S. and Xiong, Y. 2016. Characterisation of multidrug-resistant Shiga toxin-producing Escherichia coli cultured from pigs in China: co-occurrence of extended-spectrum beta-lactamase- and mcr-1-encoding genes on plasmids. *Int J Antimicrob Agents.* 48, 445-448.

Batchelor, M., Clifton-Hadley, F. A., Stallwood, A. D., Paiba, G. A., Davies, R. H. and Liebana, E. 2005. Detection of multiple cephalosporin-resistant Escherichia coli from a cattle fecal sample in Great Britain. *Microb Drug Resist.* 11, 58-61.

Belmahdi, M., Bakour, S., Al Bayssari, C., Touati, A. and Rolain, J. M. 2016. Molecular characterisation of extended-spectrum beta-lactamase- and plasmid AmpC-producing Escherichia coli strains isolated from broilers in Bejaia, Algeria. *J Glob Antimicrob Resist*. 6, 108-112.

Ben-Ami, R., Rodriguez-Bano, J., Arslan, H., Pitout, J. D., Quentin, C., Calbo, E. S., Azap, O. K., Arpin, C., Pascual, A., Livermore, D. M., Garau, J. and Carmeli, Y. 2009. A multinational survey of risk factors for infection with extended-spectrum beta-lactamase-producing enterobacteriaceae in nonhospitalized patients. *Clin Infect Dis*. 49, 682-690.

Ben Sallem, R., Ben Slama, K., Rojo-Bezares, B., Porres-Osante, N., Jouini, A., Klibi, N., Boudabous, A., Saenz, Y. and Torres, C. 2014. IncI1 plasmids carrying bla(CTX-M-1) or bla(CMY-2) genes in Escherichia coli from healthy humans and animals in Tunisia. *Microb Drug Resist*. 20, 495-500.

Ben Sallem, R., Ben Slama, K., Saenz, Y., Rojo-Bezares, B., Estepa, V., Jouini, A., Gharsa, H., Klibi, N., Boudabous, A. and Torres, C. 2012. Prevalence and characterization of extended-spectrum beta-lactamase (ESBL)- and CMY-2-producing Escherichia coli isolates from healthy food-producing animals in Tunisia. *Foodborne Pathog Dis*. 9, 1137-1142.

Ben Slama, K., Jouini, A., Ben Sallem, R., Somalo, S., Saenz, Y., Estepa, V., Boudabous, A. and Torres, C. 2010. Prevalence of broad-spectrum cephalosporin-resistant Escherichia coli isolates in food samples in Tunisia, and characterization of integrons and antimicrobial resistance mechanisms implicated. *Int J Food Microbiol*. 137, 281-286.

Bergenholtz, R. D., Jorgensen, M. S., Hansen, L. H., Jensen, L. B. and Hasman, H. 2009. Characterization of genetic determinants of extended-spectrum cephalosporinases (ESCs) in Escherichia coli isolates from Danish and imported poultry meat. *J Antimicrob Chemother*. 64, 207-209.

Bertrand, S., Weill, F. X., Cloeckaert, A., Vrints, M., Mairiaux, E., Praud, K., Dierick, K., Wildemauve, C., Godard, C., Butaye, P., Imberechts, H., Grimont, P. A. and Collard, J. M. 2006. Clonal emergence of extended-spectrum beta-lactamase (CTX-M-2)-producing Salmonella enterica serovar Virchow isolates with reduced susceptibilities to ciprofloxacin among poultry and humans in Belgium and France (2000 to 2003). *J Clin Microbiol.* 44, 2897-2903.

Blaak, H., Hamidjaja, R. A., Van Hoek, A. H., De Heer, L., De Roda Husman, A. M. and Schets, F. M. 2014. Detection of extended-spectrum beta-lactamase (ESBL)-producing Escherichia coli on flies at poultry farms. *Appl Environ Microbiol.* 80, 239-246.

Blaak, H., Van Hoek, A. H., Hamidjaja, R. A., Van Der Plaats, R. Q., Kerkhof-De Heer, L., De Roda Husman, A. M. and Schets, F. M. 2015. Distribution, Numbers, and Diversity of ESBL-Producing E. coli in the Poultry Farm Environment. *PLoS One.* 10, e0135402.

Blanc, V., Mesa, R., Saco, M., Lavilla, S., Prats, G., Miro, E., Navarro, F., Cortes, P. and Llagostera, M. 2006. ESBL- and plasmidic class C beta-lactamase-producing E. coli strains isolated from poultry, pig and rabbit farms. *Vet Microbiol.* 118, 299-304.

Bonnedahl, J., Drobni, P., Johansson, A., Hernandez, J., Melhus, A., Stedt, J., Olsen, B. and Drobni, M. 2010. Characterization, and comparison, of human clinical and black-headed gull (Larus ridibundus) extended-spectrum beta-lactamase-producing bacterial isolates from Kalmar, on the southeast coast of Sweden. *J Antimicrob Chemother.* 65, 1939-1944.

Borjesson, S., Egervarn, M., Lindblad, M. and Englund, S. 2013. Frequent occurrence of extended-spectrum beta-lactamase- and transferable ampc beta-lactamase-producing Escherichia coli on domestic chicken meat in Sweden. *Appl Environ Microbiol.* 79, 2463-2466.

Bortolaia, V., Guardabassi, L., Bisgaard, M., Larsen, J. and Bojesen, A. M. 2010a. Escherichia coli producing CTX-M-1, -2, and -9 group beta-lactamases in organic chicken egg production. *Antimicrob Agents Chemother.* 54, 3527-3528.

Bortolaia, V., Guardabassi, L., Trevisani, M., Bisgaard, M., Venturi, L. and Bojesen, A. M. 2010b. High diversity of extended-spectrum beta-lactamases in Escherichia coli isolates from Italian broiler flocks. *Antimicrob Agents Chemother.* 54, 1623-1626.

Bortolaia, V., Larsen, J., Damborg, P. and Guardabassi, L. 2011. Potential pathogenicity and host range of extended-spectrum beta-lactamase-producing Escherichia coli isolates from healthy poultry. *Appl Environ Microbiol.* 77, 5830-5833.

Boyle, F., Morris, D., O'connor, J., Delappe, N., Ward, J. and Cormican, M. 2010. First report of extended-spectrum-beta-lactamase-producing Salmonella enterica serovar Kentucky isolated from poultry in Ireland. *Antimicrob Agents Chemother.* 54, 551-553.

Bradford, P. A., Petersen, P. J., Fingerman, I. M. and White, D. G. 1999. Characterization of expanded-spectrum cephalosporin resistance in E. coli isolates associated with bovine calf diarrhoeal disease. *J Antimicrob Chemother.* 44, 607-610.

Brauer, A., Telling, K., Laht, M., Kalmus, P., Lutsar, I., Remm, M., Kisand, V. and Tenson, T. 2016. Plasmid with Colistin Resistance Gene mcr-1 in Extended-Spectrum-beta-Lactamase-Producing Escherichia coli Strains Isolated from Pig Slurry in Estonia. *Antimicrob Agents Chemother.* 60, 6933-6936.

Braun, S. D., Ahmed, M. F., El-Adawy, H., Hotzel, H., Engelmann, I., Weiss, D., Monecke, S. and Ehricht, R. 2016. Surveillance of Extended-Spectrum Beta-Lactamase-Producing Escherichia coli in Dairy Cattle Farms in the Nile Delta, Egypt. *Front Microbiol.* 7, 1020.

Brinas, L., Moreno, M. A., Teshager, T., Saenz, Y., Porrero, M. C., Dominguez, L. and Torres, C. 2005. Monitoring and characterization of extended-spectrum beta-lactamases in Escherichia coli strains from healthy and sick animals in Spain in 2003. *Antimicrob Agents Chemother.* 49, 1262-1264.

Brinas, L., Moreno, M. A., Zarazaga, M., Porrero, C., Saenz, Y., Garcia, M., Dominguez, L. and Torres, C. 2003. Detection of CMY-2, CTX-M-14, and SHV-12 beta-lactamases in Escherichia coli fecal-sample isolates from healthy chickens. *Antimicrob Agents Chemother.* 47, 2056-2058.

Brower, C. H., Mandal, S., Hayer, S., Sran, M., Zehra, A., Patel, S. J., Kaur, R., Chatterjee, L., Mishra, S., Das, B. R., Singh, P., Singh, R., Gill, J. P. S. and Laxminarayan, R. 2017. The Prevalence of Extended-Spectrum Beta-Lactamase-Producing Multidrug-Resistant Escherichia Coli in

Poultry Chickens and Variation According to Farming Practices in Punjab, India. *Environ Health Perspect.* 125, 077015.

Canton, R., Novais, A., Valverde, A., Machado, E., Peixe, L., Baquero, F. and Coque, T. M. 2008. Prevalence and spread of extended-spectrum beta-lactamase-producing Enterobacteriaceae in Europe. *Clin Microbiol Infect.* 14 Suppl 1, 144-153.

Carattoli, A. 2008. Animal reservoirs for extended spectrum beta-lactamase producers. *Clin Microbiol Infect.* 14 Suppl 1, 117-123.

Carattoli, A. 2009. Resistance plasmid families in Enterobacteriaceae. *Antimicrob Agents Chemother.* 53, 2227-2238.

Cardinale, E., Colbachini, P., Perrier-Gros-Claude, J. D., Gassama, A. and Aidara-Kane, A. 2001. Dual emergence in food and humans of a novel multiresistant serotype of Salmonella in Senegal: Salmonella enterica subsp. enterica serotype 35:c:1,2. *J Clin Microbiol.* 39, 2373-2374.

Carmo, L. P., Nielsen, L. R., Da Costa, P. M. and Alban, L. 2014. Exposure assessment of extended-spectrum beta-lactamases/AmpC beta-lactamases-producing Escherichia coli in meat in Denmark. *Infect Ecol Epidemiol.* 4.

Carneiro, C., Araujo, C., Goncalves, A., Vinue, L., Somalo, S., Ruiz, E., Uliyakina, I., Rodrigues, J., Igrejas, G., Poeta, P. and Torres, C. 2010. Detection of CTX-M-14 and TEM-52 extended-spectrum beta-lactamases in fecal Escherichia coli isolates of captive ostrich in Portugal. *Foodborne Pathog Dis.* 7, 991-994.

Cartelle, M., Del Mar Tomas, M., Molina, F., Moure, R., Villanueva, R. and Bou, G. 2004. High-level resistance to ceftazidime conferred by a novel enzyme, CTX-M-32, derived from CTX-M-1 through a single Asp240-Gly substitution. *Antimicrob Agents Chemother.* 48, 2308-2313.

Castellanos, L. R., Donado-Godoy, P., Leon, M., Clavijo, V., Arevalo, A., Bernal, J. F., Timmerman, A. J., Mevius, D. J., Wagenaar, J. A. and Hordijk, J. 2017. High Heterogeneity of Escherichia coli Sequence Types Harbouring ESBL/AmpC Genes on IncI1 Plasmids in the Colombian Poultry Chain. *PLoS One.* 12, e0170777.

Changkaew, K., Intarapuk, A., Utrarachkij, F., Nakajima, C., Suthienkul, O. and Suzuki, Y. 2015. Antimicrobial Resistance, Extended-Spectrum beta-Lactamase Productivity, and Class 1 Integrons in Escherichia coli from Healthy Swine. *J Food Prot.* 78, 1442-1450.

Chiaretto, G., Zavagnin, P., Bettini, F., Mancin, M., Minorello, C., Saccardin, C. and Ricci, A. 2008. Extended spectrum beta-lactamase SHV-12-producing Salmonella from poultry. *Vet Microbiol.* 128, 406-413.

Chishimba, K., Hang'ombe, B. M., Muzandu, K., Mshana, S. E., Matee, M. I., Nakajima, C. and Suzuki, Y. 2016. Detection of Extended-Spectrum Beta-Lactamase-Producing Escherichia coli in Market-Ready Chickens in Zambia. *Int J Microbiol.* 2016, 5275724.

Choi, D., Chon, J. W., Kim, H. S., Kim, D. H., Lim, J. S., Yim, J. H. and Seo, K. H. 2015. Incidence, Antimicrobial Resistance, and Molecular Characteristics of Nontyphoidal Salmonella Including Extended-Spectrum beta-Lactamase Producers in Retail Chicken Meat. *J Food Prot.* 78, 1932-1937.

Chon, J. W., Jung, H. I., Kuk, M., Kim, Y. J., Seo, K. H. and Kim, S. K. 2015. High occurrence of extended-spectrum beta-lactamase-producing Salmonella in broiler carcasses from poultry slaughterhouses in South Korea. *Foodborne Pathog Dis.* 12, 190-196.

Chong, Y., Ito, Y. and Kamimura, T. 2011. Genetic evolution and clinical impact in extended-spectrum beta-lactamase-producing Escherichia coli and Klebsiella pneumoniae. *Infect Genet Evol.* 11, 1499-1504.

Cloeckaert, A., Praud, K., Doublet, B., Bertini, A., Carattoli, A., Butaye, P., Imberechts, H., Bertrand, S., Collard, J. M., Arlet, G. and Weill, F. X. 2007. Dissemination of an extended-spectrum-beta-lactamase blaTEM-52 gene-carrying IncI1 plasmid in various Salmonella enterica serovars isolated from poultry and humans in Belgium and France between 2001 and 2005. *Antimicrob Agents Chemother.* 51, 1872-1875.

Cloeckaert, A., Praud, K., Lefevre, M., Doublet, B., Pardos, M., Granier, S. A., Brisabois, A. and Weill, F. X. 2010. IncI1 plasmid carrying extended-spectrum-beta-lactamase gene blaCTX-M-1 in Salmonella enterica isolates from poultry and humans in France, 2003 to 2008. *Antimicrob Agents Chemother.* 54, 4484-4486.

Cohen Stuart, J., Van Den Munckhof, T., Voets, G., Scharringa, J., Fluit, A. and Hall, M. L. 2012. Comparison of ESBL contamination in organic and conventional retail chicken meat. *Int J Food Microbiol.* 154, 212-214.

Control, E.C.F.D.P.A. 2011. Annual report of the European Anti-microbial Resistance Surveillance Network (EARS-Net). European Centre for Disease Prevention and Control.

Cortes, P., Blanc, V., Mora, A., Dahbi, G., Blanco, J. E., Blanco, M., Lopez, C., Andreu, A., Navarro, F., Alonso, M. P., Bou, G., Blanco, J. and Llagostera, M. 2010. Isolation and characterization of potentially pathogenic antimicrobial-resistant Escherichia coli strains from chicken and pig farms in Spain. *Appl Environ Microbiol.* 76, 2799-2805.

Costa, D., Vinue, L., Poeta, P., Coelho, A. C., Matos, M., Saenz, Y., Somalo, S., Zarazaga, M., Rodrigues, J. and Torres, C. 2009. Prevalence of extended-spectrum beta-lactamase-producing Escherichia coli isolates in faecal samples of broilers. *Vet Microbiol.* 138, 339-344.

Cottell, J. L., Webber, M. A., Coldham, N. G., Taylor, D. L., Cerdeno-Tarraga, A. M., Hauser, H., Thomson, N. R., Woodward, M. J. and Piddock, L. J. 2011. Complete sequence and molecular epidemiology of IncK epidemic plasmid encoding blaCTX-M-14. *Emerg Infect Dis.* 17, 645-652.

Daehre, K., Projahn, M., Semmler, T., Roesler, U. and Friese, A. 2017. Extended-Spectrum Beta-Lactamase-/AmpC Beta-Lactamase-Producing Enterobacteriaceae in Broiler Farms: Transmission Dynamics at Farm Level. *Microb Drug Resist.*

Dahmen, S., Metayer, V., Gay, E., Madec, J. Y. and Haenni, M. 2013. Characterization of extended-spectrum beta-lactamase (ESBL)-carrying plasmids and clones of Enterobacteriaceae causing cattle mastitis in France. *Vet Microbiol.* 162, 793-799.

Dahms, C., Hubner, N. O., Kossow, A., Mellmann, A., Dittmann, K. and Kramer, A. 2015. Occurrence of ESBL-Producing Escherichia coli in Livestock and Farm Workers in Mecklenburg-Western Pomerania, Germany. *PLoS One.* 10, e0143326.

DANMAP. 2010. Use of antimicrobial agents and occurrence of antimicrobial resistance inbacteria from food animals, foods and humans in Denmark. The Danish Integrated Antimicrobial resistance Monitoring and Research Program.

DANMAP. 2015. Use of antimicrobial agents and occurrence of antimicrobial resistance inbacteria from food animals, foods and humans in Denmark. The Danish Integrated Antimicrobial Resistance Monitoring and Research Program.

De Jong, A., Smet, A., Ludwig, C., Stephan, B., De Graef, E., Vanrobaeys, M. and Haesebrouck, F. 2014. Antimicrobial susceptibility of Salmonella isolates from healthy pigs and chickens (2008-2011). *Vet Microbiol.* 171, 298-306.

DEFRA. 2013. Extended-Spectrum Beta-Lactamases (ESBL) in bacteria associated with animals. (F.a.R.A. Department for Environment, ed.).

Dhanji, H., Murphy, N. M., Doumith, M., Durmus, S., Lee, S. S., Hope, R., Woodford, N. and Livermore, D. M. 2010. Cephalosporin resistance mechanisms in Escherichia coli isolated from raw chicken imported into the UK. *J Antimicrob Chemother.* 65, 2534-2537.

Dierikx, C., Van Der Goot, J., Fabri, T., Van Essen-Zandbergen, A., Smith, H. and Mevius, D. 2013a. Extended-spectrum-beta-lactamase- and AmpC-beta-lactamase-producing Escherichia coli in Dutch broilers and broiler farmers. *J Antimicrob Chemother.* 68, 60-67.

Dierikx, C., Van Essen-Zandbergen, A., Veldman, K., Smith, H. and Mevius, D. 2010a. Increased detection of extended spectrum beta-lactamase producing Salmonella enterica and Escherichia coli isolates from poultry. *Vet Microbiol.* 145, 273-278.

Dierikx, C. M., Fabri, T. and Van Der Goot, J. A. 2010b. Prevalence of extendedspectrum-β-lactamase producing E. coli isolates on broiler farms in the Netherlands. *Dutch J Med Microbiol.* S28–S29.

Dierikx, C. M., Van Der Goot, J. A., Smith, H. E., Kant, A. and Mevius, D. J. 2013b. Presence of ESBL/AmpC-producing Escherichia coli in the broiler production pyramid: a descriptive study. *PLoS One.* 8, e79005.

Djeffal, S., Bakour, S., Mamache, B., Elgroud, R., Agabou, A., Chabou, S., Hireche, S., Bouaziz, O., Rahal, K. and Rolain, J. M. 2017. Prevalence and clonal relationship of ESBL-producing Salmonella strains from humans and poultry in northeastern Algeria. *BMC Vet Res.* 13, 132.

Dohmen, W., Bonten, M. J., Bos, M. E., Van Marm, S., Scharringa, J., Wagenaar, J. A. and Heederik, D. J. 2015. Carriage of extended-spectrum beta-lactamases in pig farmers is associated with occurrence in pigs. *Clin Microbiol Infect.* 21, 917-923.

Dohmen, W., Dorado-Garcia, A., Bonten, M. J., Wagenaar, J. A., Mevius, D. and Heederik, D. J. 2017a. Risk factors for ESBL-producing Escherichia coli on pig farms: A longitudinal study in the context of reduced use of antimicrobials. *PLoS One.* 12, e0174094.

Dohmen, W., Schmitt, H., Bonten, M. and Heederik, D. 2017b. Air exposure as a possible route for ESBL in pig farmers. *Environ Res.* 155, 359-364.

Doi, Y., Paterson, D. L., Egea, P., Pascual, A., Lopez-Cerero, L., Navarro, M. D., Adams-Haduch, J. M., Qureshi, Z. A., Sidjabat, H. E. and Rodriguez-Bano, J. 2010. Extended-spectrum and CMY-type beta-lactamase-producing Escherichia coli in clinical samples and retail meat from Pittsburgh, USA and Seville, Spain. *Clin Microbiol Infect.* 16, 33-38.

Donaldson, S. C., Straley, B. A., Hegde, N. V., Sawant, A. A., Debroy, C. and Jayarao, B. M. 2006. Molecular epidemiology of ceftiofur-resistant Escherichia coli isolates from dairy calves. *Appl Environ Microbiol.* 72, 3940-3948.

Duan, R. S., Sit, T. H., Wong, S. S., Wong, R. C., Chow, K. H., Mak, G. C., Yam, W. C., Ng, L. T., Yuen, K. Y. and Ho, P. L. 2006. Escherichia coli producing CTX-M beta-lactamases in food animals in Hong Kong. *Microb Drug Resist.* 12, 145-148.

Dutil, L., Irwin, R., Finley, R., Ng, L. K., Avery, B., Boerlin, P., Bourgault, A. M., Cole, L., Daignault, D., Desruisseau, A., Demczuk, W., Hoang, L., Horsman, G. B., Ismail, J., Jamieson, F., Maki, A., Pacagnella, A. and Pillai, D. R. 2010. Ceftiofur resistance in Salmonella enterica serovar Heidelberg from chicken meat and humans, Canada. *Emerg Infect Dis.* 16, 48-54.

Egea, P., Lopez-Cerero, L., Torres, E., Gomez-Sanchez Mdel, C., Serrano, L., Navarro Sanchez-Ortiz, M. D., Rodriguez-Bano, J. and Pascual, A.

2012. Increased raw poultry meat colonization by extended spectrum beta-lactamase-producing Escherichia coli in the south of Spain. *Int J Food Microbiol.* 159, 69-73.

Egervarn, M., Borjesson, S., Byfors, S., Finn, M., Kaipe, C., Englund, S. and Lindblad, M. 2014. Escherichia coli with extended-spectrum beta-lactamases or transferable AmpC beta-lactamases and Salmonella on meat imported into Sweden. *Int J Food Microbiol.* 171, 8-14.

Eisenberger, D., Carl, A., Balsliemke, J., Kampf, P., Nickel, S., Schulze, G. and Valenza, G. 2018. Molecular Characterization of Extended-Spectrum beta-Lactamase-Producing Escherichia coli Isolates from Milk Samples of Dairy Cows with Mastitis in Bavaria, Germany. *Microb Drug Resist.* 24, 505-510.

El-Shazly, D. A., Nasef, S. A., Mahmoud, F. F. and Jonas, D. 2017. Expanded spectrum beta-lactamase producing Escherichia coli isolated from chickens with colibacillosis in Egypt. *Poult Sci.* 96, 2375-2384.

EMA. 2014. European surveillance of veterinary antimicrobial consumption. Sales of veterinary antimicrobial agents in 26 EU/EEA countries in 2012. In: *Fourth ESVAC report*, European Medicines Agency.

Endimiani, A., Bertschy, I. and Perreten, V. 2012a. Escherichia coli producing CMY-2 beta-lactamase in bovine mastitis milk. *J Food Prot.* 75, 137-138.

Endimiani, A., Hilty, M. and Perreten, V. 2012b. CMY-2-producing Escherichia coli in the nose of pigs. *Antimicrob Agents Chemother.* 56, 4556-4557.

Endimiani, A., Rossano, A., Kunz, D., Overesch, G. and Perreten, V. 2012c. First countrywide survey of third-generation cephalosporin-resistant Escherichia coli from broilers, swine, and cattle in Switzerland. *Diagn Microbiol Infect Dis.* 73, 31-38.

Escudero, E., Vinue, L., Teshager, T., Torres, C. and Moreno, M. A. 2010. Resistance mechanisms and farm-level distribution of fecal Escherichia coli isolates resistant to extended-spectrum cephalosporins in pigs in Spain. *Res Vet Sci.* 88, 83-87.

European Food Safety Authority, E.C.F.D.P.A.C. 2011. EU summary report on antimicrobial resistance in zoonotic and indicator bacteria from humans, animals and food 2009. *EFSA Journal.* 9, 2154.

European Food Safety Authority, E.C.F.D.P.A.C. 2017. The European Union summary report on antimicrobial resistance in zoonotic and indicator bacteria from humans, animals and food in 2015. *EFSA Journal.* 15, e04694.

FDA. 2014. National Antimicrobial Resistance Monitoring System—Enteric Bacteria (NARMS): 2014 Executive Report.

Ferreira, J. C., Penha Filho, R. A., Andrade, L. N., Berchieri, A., Jr. and Darini, A. L. 2014. Detection of chromosomal bla(CTX-M-2) in diverse Escherichia coli isolates from healthy broiler chickens. *Clin Microbiol Infect.* 20, O623-626.

Ferreira, J. C., Penha Filho, R. A., Andrade, L. N., Berchieri Junior, A. and Darini, A. L. 2016. Evaluation and characterization of plasmids carrying CTX-M genes in a non-clonal population of multidrug-resistant Enterobacteriaceae isolated from poultry in Brazil. *Diagn Microbiol Infect Dis.* 85, 444-448.

Fey, P. D., Safranek, T. J., Rupp, M. E., Dunne, E. F., Ribot, E., Iwen, P. C., Bradford, P. A., Angulo, F. J. and Hinrichs, S. H. 2000. Ceftriaxone-resistant salmonella infection acquired by a child from cattle. *N Engl J Med.* 342, 1242-1249.

Fischer, J., Hille, K., Ruddat, I., Mellmann, A., Kock, R. and Kreienbrock, L. 2017. Simultaneous occurrence of MRSA and ESBL-producing Enterobacteriaceae on pig farms and in nasal and stool samples from farmers. *Vet Microbiol.* 200, 107-113.

Franco, A., Leekitcharoenphon, P., Feltrin, F., Alba, P., Cordaro, G., Iurescia, M., Tolli, R., D'incau, M., Staffolani, M., Di Giannatale, E., Hendriksen, R. S. and Battisti, A. 2015. Emergence of a Clonal Lineage of Multidrug-Resistant ESBL-Producing Salmonella Infantis Transmitted from Broilers and Broiler Meat to Humans in Italy between 2011 and 2014. *PLoS One.* 10, e0144802.

Freire Martin, I., Abuoun, M., Reichel, R., La Ragione, R. M. and Woodward, M. J. 2014. Sequence analysis of a CTX-M-1 IncI1 plasmid found in Salmonella 4,5,12:i:-, Escherichia coli and Klebsiella pneumoniae on a UK pig farm. *J Antimicrob Chemother.* 69, 2098-2101.

Freitag, C., Michael, G. B., Kadlec, K., Hassel, M. and Schwarz, S. 2017. Detection of plasmid-borne extended-spectrum beta-lactamase (ESBL) genes in Escherichia coli isolates from bovine mastitis. *Vet Microbiol.* 200, 151-156.

Frye, J. G., Fedorka-Cray, P. J., Jackson, C. R. and Rose, M. 2008. Analysis of Salmonella enterica with reduced susceptibility to the third-generation cephalosporin ceftriaxone isolated from U.S. cattle during 2000-2004. *Microb Drug Resist.* 14, 251-258.

Gao, L., Hu, J., Zhang, X., Ma, R., Gao, J., Li, S., Zhao, M., Miao, Z. and Chai, T. 2014. Dissemination of ESBL-producing Escherichia coli of chicken origin to the nearby river water. *J Mol Microbiol Biotechnol.* 24, 279-285.

Gao, L., Hu, J., Zhang, X., Wei, L., Li, S., Miao, Z. and Chai, T. 2015a. Application of swine manure on agricultural fields contributes to extended-spectrum beta-lactamase-producing Escherichia coli spread in Tai'an, China. *Front Microbiol.* 6, 313.

Gao, L., Tan, Y., Zhang, X., Hu, J., Miao, Z., Wei, L. and Chai, T. 2015b. Emissions of Escherichia coli carrying extended-spectrum beta-lactamase resistance from pig farms to the surrounding environment. *Int J Environ Res Public Health.* 12, 4203-4213.

Geser, N., Stephan, R. and Hachler, H. 2012. Occurrence and characteristics of extended-spectrum beta-lactamase (ESBL) producing Enterobacteriaceae in food producing animals, minced meat and raw milk. *BMC Vet Res.* 8, 21.

Geser, N., Stephan, R., Kuhnert, P., Zbinden, R., Kaeppeli, U., Cernela, N. and Haechler, H. 2011. Fecal carriage of extended-spectrum beta-lactamase-producing Enterobacteriaceae in swine and cattle at slaughter in Switzerland. *J Food Prot.* 74, 446-449.

Ghodousi, A., Bonura, C., Di Noto, A. M. and Mammina, C. 2015. Extended-Spectrum ss-Lactamase, AmpC-Producing, and Fluoroquinolone-Resistant Escherichia coli in Retail Broiler Chicken Meat, Italy. *Foodborne Pathog Dis.* 12, 619-625.

Girlich, D., Poirel, L., Carattoli, A., Kempf, I., Lartigue, M. F., Bertini, A. and Nordmann, P. 2007. Extended-spectrum beta-lactamase CTX-M-1 in Escherichia coli isolates from healthy poultry in France. *Appl Environ Microbiol.* 73, 4681-4685.

Giufre, M., Graziani, C., Accogli, M., Luzzi, I., Busani, L., Cerquetti, M. and Escherichia Coli Study, G. 2012. Escherichia coli of human and avian origin: detection of clonal groups associated with fluoroquinolone and multidrug resistance in Italy. *J Antimicrob Chemother.* 67, 860-867.

Goncalves, A., Torres, C., Silva, N., Carneiro, C., Radhouani, H., Coelho, C., Araujo, C., Rodrigues, J., Vinue, L., Somalo, S., Poeta, P. and Igrejas, G. 2010. Genetic characterization of extended-spectrum beta-lactamases in Escherichia coli isolates of pigs from a Portuguese intensive swine farm. *Foodborne Pathog Dis.* 7, 1569-1573.

Gonggrijp, M. A., Santman-Berends, I., Heuvelink, A. E., Buter, G. J., Van Schaik, G., Hage, J. J. and Lam, T. 2016. Prevalence and risk factors for extended-spectrum beta-lactamase- and AmpC-producing Escherichia coli in dairy farms. *J Dairy Sci.* 99, 9001-9013.

Grami, R., Dahmen, S., Mansour, W., Mehri, W., Haenni, M., Aouni, M. and Madec, J. Y. 2014. blaCTX-M-15-carrying F2:A-:B- plasmid in Escherichia coli from cattle milk in Tunisia. *Microb Drug Resist.* 20, 344-349.

Grami, R., Mansour, W., Dahmen, S., Mehri, W., Haenni, M., Aouni, M. and Madec, J. Y. 2013. The blaCTX-M-1 IncI1/ST3 plasmid is dominant in chickens and pets in Tunisia. *J Antimicrob Chemother.* 68, 2950-2952.

Gundogan, N., Citak, S. and Yalcin, E. 2011. Virulence properties of extended spectrum beta-lactamase-producing Klebsiella species in meat samples. *J Food Prot.* 74, 559-564.

Guo, Y. F., Zhang, W. H., Ren, S. Q., Yang, L., Lu, D. H., Zeng, Z. L., Liu, Y. H. and Jiang, H. X. 2014. IncA/C plasmid-mediated spread of CMY-2 in multidrug-resistant Escherichia coli from food animals in China. *PLoS One.* 9, e96738.

Gupta, A., Fontana, J., Crowe, C., Bolstorff, B., Stout, A., Van Duyne, S., Hoekstra, M. P., Whichard, J. M., Barrett, T. J., Angulo, F. J. and

National Antimicrobial Resistance Monitoring System Pulsenet Working, G. 2003. Emergence of multidrug-resistant Salmonella enterica serotype Newport infections resistant to expanded-spectrum cephalosporins in the United States. *J Infect Dis.* 188, 1707-1716.

Gutkind, G. O., Di Conza, J., Power, P. and Radice, M. 2013. beta-lactamase-mediated resistance: a biochemical, epidemiological and genetic overview. *Curr Pharm Des.* 19, 164-208.

Hachler, H., Kotsakis, S. D., Tzouvelekis, L. S., Geser, N., Lehner, A., Miriagou, V. and Stephan, R. 2013. Characterisation of CTX-M-117, a Pro174Gln variant of CTX-M-15 extended-spectrum beta-lactamase, from a bovine Escherichia coli isolate. *Int J Antimicrob Agents.* 41, 94-95.

Hammerum, A. M., Larsen, J., Andersen, V. D., Lester, C. H., Skovgaard Skytte, T. S., Hansen, F., Olsen, S. S., Mordhorst, H., Skov, R. L., Aarestrup, F. M. and Agerso, Y. 2014. Characterization of extended-spectrum beta-lactamase (ESBL)-producing Escherichia coli obtained from Danish pigs, pig farmers and their families from farms with high or no consumption of third- or fourth-generation cephalosporins. *J Antimicrob Chemother.* 69, 2650-2657.

Han, J. W., Koh, H. B. and Kim, T. J. 2016. Molecular Characterization of beta-Lactamase-Producing Escherichia coli Collected from 2001 to 2011 from Pigs in Korea. *Foodborne Pathog Dis.* 13, 68-76.

Hansen, K. H., Bortolaia, V., Nielsen, C. A., Nielsen, J. B., Schonning, K., Agerso, Y. and Guardabassi, L. 2016. Host-Specific Patterns of Genetic Diversity among IncI1-Igamma and IncK Plasmids Encoding CMY-2 beta-Lactamase in Escherichia coli Isolates from Humans, Poultry Meat, Poultry, and Dogs in Denmark. *Appl Environ Microbiol.* 82, 4705-4714.

Harris, A. D., Mcgregor, J. C., Johnson, J. A., Strauss, S. M., Moore, A. C., Standiford, H. C., Hebden, J. N. and Morris, J. G., JR. 2007. Risk factors for colonization with extended-spectrum beta-lactamase-producing bacteria and intensive care unit admission. *Emerg Infect Dis.* 13, 1144-1149.

Hartmann, A., Locatelli, A., Amoureux, L., Depret, G., Jolivet, C., Gueneau, E. and Neuwirth, C. 2012. Occurrence of CTX-M Producing

Escherichia coli in Soils, Cattle, and Farm Environment in France (Burgundy Region). *Front Microbiol.* 3, 83.

Hasan, B., Islam, K., Ahsan, M., Hossain, Z., Rashid, M., Talukder, B., Ahmed, K. U., Olsen, B. and Abul Kashem, M. 2014. Fecal carriage of multi-drug resistant and extended spectrum beta-lactamases producing E. coli in household pigeons, Bangladesh. *Vet Microbiol.* 168, 221-224.

Hasan, B., Sandegren, L., Melhus, A., Drobni, M., Hernandez, J., Waldenstrom, J., Alam, M. and Olsen, B. 2012. Antimicrobial drug-resistant Escherichia coli in wild birds and free-range poultry, Bangladesh. *Emerg Infect Dis.* 18, 2055-2058.

Hasman, H., Hammerum, A. M., Hansen, F., Hendriksen, R. S., Olesen, B., Agerso, Y., Zankari, E., Leekitcharoenphon, P., Stegger, M., Kaas, R. S., Cavaco, L. M., Hansen, D. S., Aarestrup, F. M. and Skov, R. L. 2015. Detection of mcr-1 encoding plasmid-mediated colistin-resistant Escherichia coli isolates from human bloodstream infection and imported chicken meat, Denmark 2015. *Euro Surveill.* 20.

Hasman, H., Mevius, D., Veldman, K., Olesen, I. and Aarestrup, F. M. 2005. beta-Lactamases among extended-spectrum beta-lactamase (ESBL)-resistant Salmonella from poultry, poultry products and human patients in The Netherlands. *J Antimicrob Chemother.* 56, 115-121.

Hawkey, P. M. and Jones, A. M. 2009. The changing epidemiology of resistance. *J Antimicrob Chemother.* 64 Suppl 1, i3-10.

Hiki, M., Usui, M., Kojima, A., Ozawa, M., Ishii, Y. and Asai, T. 2013. Diversity of plasmid replicons encoding the bla(CMY-2) gene in broad-spectrum cephalosporin-resistant Escherichia coli from livestock animals in Japan. *Foodborne Pathog Dis.* 10, 243-249.

Hindermann, D., Gopinath, G., Chase, H., Negrete, F., Althaus, D., Zurfluh, K., Tall, B. D., Stephan, R. and Nuesch-Inderbinen, M. 2017. Salmonella enterica serovar Infantis from Food and Human Infections, Switzerland, 2010-2015: Poultry-Related Multidrug Resistant Clones and an Emerging ESBL Producing Clonal Lineage. *Front Microbiol.* 8, 1322.

Hiroi, M., Harada, T., Kawamori, F., Takahashi, N., Kanda, T., Sugiyama, K., Masuda, T., Yoshikawa, Y. and Ohashi, N. 2011. A survey of beta-lactamase-producing Escherichia coli in farm animals and raw retail meat in Shizuoka Prefecture, Japan. *Jpn J Infect Dis.* 64, 153-155.

Hiroi, M., Yamazaki, F., Harada, T., Takahashi, N., Iida, N., Noda, Y., Yagi, M., Nishio, T., Kanda, T., Kawamori, F., Sugiyama, K., Masuda, T., Hara-Kudo, Y. and Ohashi, N. 2012. Prevalence of extended-spectrum beta-lactamase-producing Escherichia coli and Klebsiella pneumoniae in food-producing animals. *J Vet Med Sci.* 74, 189-195.

Ho, P. L., Chow, K. H., Lai, E. L., Lo, W. U., Yeung, M .K., Chan, J., Chan, P. Y. and Yuen, K. Y. 2011. Extensive dissemination of CTX-M-producing Escherichia coli with multidrug resistance to 'critically important' antibiotics among food animals in Hong Kong, 2008-10. *J Antimicrob Chemother.* 66, 765-768.

Hopkins, K. L., Liebana, E., Villa, L., Batchelor, M., Threlfall, E. J. and Carattoli, A. 2006. Replicon typing of plasmids carrying CTX-M or CMY beta-lactamases circulating among Salmonella and Escherichia coli isolates. *Antimicrob Agents Chemother.* 50, 3203-3206.

Horton, R. A., Duncan, D., Randall, L. P., Chappell, S., Brunton, L. A., Warner, R., Coldham, N. G. and Teale, C. J. 2016. Longitudinal study of CTX-M ESBL-producing E. coli strains on a UK dairy farm. *Res Vet Sci.* 109, 107-113.

Horton, R. A., Randall, L. P., Snary, E. L., Cockrem, H., Lotz, S., Wearing, H., Duncan, D., Rabie, A., Mclaren, I., Watson, E., La Ragione, R. M. and Coldham, N. G. 2011. Fecal carriage and shedding density of CTX-M extended-spectrum {beta}-lactamase-producing escherichia coli in cattle, chickens, and pigs: implications for environmental contamination and food production. *Appl Environ Microbiol.* 77, 3715-3719.

Hu, G. Z., Chen, H. Y., Si, H. B., Deng, L. X., Wei, Z. Y., Yuan, L. and Kuang, X. H. 2008. Phenotypic and molecular characterization of TEM-116 extended-spectrum beta-lactamase produced by a Shigella flexneri clinical isolate from chickens. *FEMS Microbiol Lett.* 279, 162-166.

Huijbers, P. M., De Kraker, M., Graat, E. A., Van Hoek, A. H., Van Santen, M. G., De Jong, M. C., Van Duijkeren, E. and De Greeff, S. C. 2013. Prevalence of extended-spectrum beta-lactamase-producing Enterobacteriaceae in humans living in municipalities with high and low broiler density. *Clin Microbiol Infect.* 19, E256-259.

Huijbers, P. M., Graat, E. A., Haenen, A. P., Van Santen, M. G., Van Essen-Zandbergen, A., Mevius, D. J., Van Duijkeren, E. and Van Hoek, A. H. 2014. Extended-spectrum and AmpC beta-lactamase-producing Escherichia coli in broilers and people living and/or working on broiler farms: prevalence, risk factors and molecular characteristics. *J Antimicrob Chemother.* 69, 2669-2675.

Huijbers, P. M., Van Hoek, A. H., Graat, E. A., Haenen, A. P., Florijn, A., Hengeveld, P. D. and Van Duijkeren, E. 2015. Methicillin-resistant Staphylococcus aureus and extended-spectrum and AmpC beta-lactamase-producing Escherichia coli in broilers and in people living and/or working on organic broiler farms. *Vet Microbiol.* 176, 120-125.

Hunter, P. A., Dawson, S., French, G. L., Goossens, H., Hawkey, P. M., Kuijper, E. J., Nathwani, D., Taylor, D. J., Teale, C. J., Warren, R. E., Wilcox, M. H., Woodford, N., Wulf, M. W. and Piddock, L. J. 2010. Antimicrobial-resistant pathogens in animals and man: prescribing, practices and policies. *J Antimicrob Chemother.* 65 Suppl 1, i3-17.

Hur, J., Kim, J. H., Park, J. H., Lee, Y. J. and Lee, J. H. 2011. Molecular and virulence characteristics of multi-drug resistant Salmonella Enteritidis strains isolated from poultry. *Vet J.* 189, 306-311.

Ibrahim, D. R., Dodd, C. E., Stekel, D. J., Ramsden, S. J. and Hobman, J. L. 2016. Multidrug resistant, extended spectrum beta-lactamase (ESBL)-producing Escherichia coli isolated from a dairy farm. *FEMS Microbiol Ecol.* 92, fiw013.

Jacoby, G. A. 2009. AmpC beta-lactamases. *Clin Microbiol Rev.* 22, 161-182, Table of Contents.

Jahanbakhsh, S., Letellier, A. and Fairbrother, J. M. 2016. Circulating of CMY-2 beta-lactamase gene in weaned pigs and their environment in a commercial farm and the effect of feed supplementation with a clay mineral. *J Appl Microbiol.* 121, 136-148.

Jakobsen, L., Bortolaia, V., Bielak, E., Moodley, A., Olsen, S. S., Hansen, D. S., Frimodt-Moller, N., Guardabassi, L. and Hasman, H. 2015. Limited similarity between plasmids encoding CTX-M-1 beta-lactamase in Escherichia coli from humans, pigs, cattle, organic poultry layers and horses in Denmark. *J Glob Antimicrob Resist.* 3, 132-136.

Jensen, L. B., Hasman, H., Agerso, Y., Emborg, H. D. and Aarestrup, F. M. 2006. First description of an oxyimino-cephalosporin-resistant, ESBL-carrying Escherichia coli isolated from meat sold in Denmark. *J Antimicrob Chemother.* 57, 793-794.

Jouini, A., Vinue, L., Slama, K. B., Saenz, Y., Klibi, N., Hammami, S., Boudabous, A. and Torres, C. 2007. Characterization of CTX-M and SHV extended-spectrum beta-lactamases and associated resistance genes in Escherichia coli strains of food samples in Tunisia. *J Antimicrob Chemother.* 60, 1137-1141.

Kameyama, M., Chuma, T., Yabata, J., Tominaga, K., Iwata, H. and Okamoto, K. 2013. Prevalence and epidemiological relationship of CMY-2 AmpC beta-lactamase and CTX-M extended-spectrum beta-lactamase-producing Escherichia coli isolates from broiler farms in Japan. *J Vet Med Sci.* 75, 1009-1015.

Kar, D., Bandyopadhyay, S., Bhattacharyya, D., Samanta, I., Mahanti, A., Nanda, P. K., Mondal, B., Dandapat, P., Das, A. K., Dutta, T. K., Bandyopadhyay, S. and Singh, R. K. 2015. Molecular and phylogenetic characterization of multidrug resistant extended spectrum beta-lactamase producing Escherichia coli isolated from poultry and cattle in Odisha, India. *Infect Genet Evol.* 29, 82-90.

Kataoka, Y., Murakami, K., Torii, Y., Kimura, H., Maeda-Mitani, E., Shigemura, H., Fujimoto, S. and Murakami, S. 2017. Reduction in the prevalence of AmpC beta-lactamase CMY-2 in Salmonella from chicken meat following cessation of the use of ceftiofur in Japan. *J Glob Antimicrob Resist.* 10, 10-11.

Kawamura, K., Goto, K., Nakane, K. and Arakawa, Y. 2014. Molecular epidemiology of extended-spectrum beta-lactamases and Escherichia coli isolated from retail foods including chicken meat in Japan. *Foodborne Pathog Dis.* 11, 104-110.

Kilani, H., Abbassi, M. S., Ferjani, S., Mansouri, R., Sghaier, S., Ben Salem, R., Jaouani, I., Douja, G., Brahim, S., Hammami, S., Ben Chehida, N. and Boubaker, I. B. 2015. Occurrence of bla CTX-M-1, qnrB1 and virulence genes in avian ESBL-producing Escherichia coli isolates from Tunisia. *Front Cell Infect Microbiol.* 5, 38.

Kim, S., Kang, H. W. and Woo, G. J. 2015. Prevalence of CTX-M-15 Extended-Spectrum beta-Lactamase-Producing Salmonella Isolated from Chicken in Korea. Foodborne *Pathog Dis.* 12, 661-663.

Kim, S. H. and Wei, C. I. 2007. Expression of AmpC beta-lactamase in Enterobacter cloacae isolated from retail ground beef, cattle farm and processing facilities. *J Appl Microbiol.* 103, 400-408.

Kim, Y., Moon, J., Oh, D., Chon, J., Song, B., Lim, J., Heo, E., Park, H., Wee, S. and Sung, K. 2018. Genotypic characterization of ESBL-producing E. coli from imported meat in South Korea. Food Research International.

King, L. A., Nogareda, F., Weill, F. X., Mariani-Kurkdjian, P., Loukiadis, E., Gault, G., Jourdan-Dasilva, N., Bingen, E., Mace, M., Thevenot, D., Ong, N., Castor, C., Noel, H., Van Cauteren, D., Charron, M., Vaillant, V., Aldabe, B., Goulet, V., Delmas, G., Couturier, E., Le Strat, Y., Combe, C., Delmas, Y., Terrier, F., Vendrely, B., Rolland, P. and De Valk, H. 2012. Outbreak of Shiga toxin-producing Escherichia coli O104:H4 associated with organic fenugreek sprouts, France, June 2011. *Clin Infect Dis.* 54, 1588-1594.

Kirchner, M., Wearing, H., Hopkins, K. L. and Teale, C. 2011. Characterization of plasmids encoding cefotaximases group 1 enzymes in Escherichia coli recovered from cattle in England and Wales. *Microb Drug Resist.* 17, 463-470.

Koga, V. L., Rodrigues, G. R., Scandorieiro, S., Vespero, E. C., Oba, A., De Brito, B. G., De Brito, K. C., Nakazato, G. and Kobayashi, R. K. 2015a. Evaluation of the Antibiotic Resistance and Virulence of Escherichia coli Strains Isolated from Chicken Carcasses in 2007 and 2013 from Parana, Brazil. *Foodborne Pathog Dis.* 12, 479-485.

Koga, V. L., Scandorieiro, S., Vespero, E. C., Oba, A., De Brito, B. G., De Brito, K. C. T., Nakazato, G. and Kobayashi, R. K. T. 2015b. Comparison of Antibiotic Resistance and Virulence Factors among Escherichia coli Isolated from Conventional and Free-Range Poultry. *BioMed Research International.* 2015.

Kojima, A., Ishii, Y., Ishihara, K., Esaki, H., Asai, T., Oda, C., Tamura, Y., Takahashi, T. and Yamaguchi, K. 2005. Extended-spectrum-beta-lactamase-producing Escherichia coli strains isolated from farm animals from 1999 to 2002: report from the Japanese Veterinary Antimicrobial Resistance Monitoring Program. *Antimicrob Agents Chemother.* 49, 3533-3537.

Kola, A., Kohler, C., Pfeifer, Y., Schwab, F., Kuhn, K., Schulz, K., Balau, V., Breitbach, K., Bast, A., Witte, W., Gastmeier, P. and Steinmetz, I. 2012. High prevalence of extended-spectrum-beta-lactamase-producing Enterobacteriaceae in organic and conventional retail chicken meat, Germany. *J Antimicrob Chemother.* 67, 2631-2634.

Kolar, M., Bardon, J., Chroma, M., Hricova, K., Stosova, T., Sauer, P. and Koukalova, D. 2010 ESBL and AmpC beta-lactamase-producing Enterobacteriaceae in poultry in the Czech Republic. *Veterinarni Medicina.* 55, 119-124.

Koovapra, S., Bandyopadhyay, S., Das, G., Bhattacharyya, D., Banerjee, J., Mahanti, A., Samanta, I., Nanda, P. K., Kumar, A., Mukherjee, R., Dimri, U. and Singh, R. K. 2016. Molecular signature of extended spectrum beta-lactamase producing Klebsiella pneumoniae isolated from bovine milk in eastern and north-eastern India. *Infect Genet Evol.* 44, 395-402.

Kozak, G. K., Boerlin, P., Janecko, N., Reid-Smith, R. J. and Jardine, C. 2009. Antimicrobial resistance in Escherichia coli isolates from swine and wild small mammals in the proximity of swine farms and in natural environments in Ontario, Canada. *Appl Environ Microbiol.* 75, 559-566.

Kraemer, J. G., Pires, J., Kueffer, M., Semaani, E., Endimiani, A., Hilty, M. and Oppliger, A. 2017. Prevalence of extended-spectrum beta-lactamase-producing Enterobacteriaceae and Methicillin-Resistant Staphylococcus aureus in pig farms in Switzerland. *Sci Total Environ.* 603-604, 401-405.

Laube, H., Friese, A., Von Salviati, C., Guerra, B., Kasbohrer, A., Kreienbrock, L. and Roesler, U. 2013. Longitudinal monitoring of extended-spectrum-beta-lactamase/AmpC-producing Escherichia coli at German broiler chicken fattening farms. *Appl Environ Microbiol.* 79, 4815-4820.

Laube, H., Friese, A., Von Salviati, C., Guerra, B. and Rosler, U. 2014. Transmission of ESBL/AmpC-producing Escherichia coli from broiler chicken farms to surrounding areas. *Vet Microbiol.* 172, 519-527.

Laupland, K. B., Church, D. L., Vidakovich, J., Mucenski, M. and Pitout, J. D. 2008. Community-onset extended-spectrum beta-lactamase (ESBL) producing Escherichia coli: importance of international travel. *J Infect.* 57, 441-448.

Lavilla, S., Gonzalez-Lopez, J. J., Miro, E., Dominguez, A., Llagostera, M., Bartolome, R. M., Mirelis, B., Navarro, F. and Prats, G. 2008. Dissemination of extended-spectrum beta-lactamase-producing bacteria: the food-borne outbreak lesson. *J Antimicrob Chemother.* 61, 1244-1251.

Lee, H. Y., Su, L. H., Tsai, M. H., Kim, S. W., Chang, H. H., Jung, S. I., Park, K. H., Perera, J., Carlos, C., Tan, B. H., Kumarasinghe, G., So, T., Chongthaleong, A., Hsueh, P. R., Liu, J. W., Song, J. H. and Chiu, C. H. 2009. High rate of reduced susceptibility to ciprofloxacin and ceftriaxone among nontyphoid Salmonella clinical isolates in Asia. *Antimicrob Agents Chemother.* 53, 2696-2699.

Lee, K. E., Lim, S. I., Choi, H. W., Lim, S. K., Song, J. Y. and An, D. J. 2014. Plasmid-mediated AmpC beta-lactamase (CMY-2) gene in Salmonella typhimurium isolated from diarrheic pigs in South Korea. *BMC Res Notes.* 7, 329.

Leverstein-Van Hall, M. A., Dierikx, C. M., Cohen Stuart, J., Voets, G. M., Van Den Munckhof, M. P., Van Essen-Zandbergen, A., Platteel, T., Fluit, A. C., Van De Sande-Bruinsma, N., Scharinga, J., Bonten, M. J., Mevius, D. J. and National, E. S. G. 2011. Dutch patients, retail chicken meat and poultry share the same ESBL genes, plasmids and strains. *Clin Microbiol Infect.* 17, 873-880.

Li, J., Ma, Y., Hu, C., Jin, S., Zhang, Q., Ding, H., Ran, L. and Cui, S. 2010a. Dissemination of cefotaxime-M-producing Escherichia coli isolates in poultry farms, but not swine farms, in China. *Foodborne Pathog Dis.* 7, 1387-1392.

Li, L., Jiang, Z. G., Xia, L. N., Shen, J. Z., Dai, L., Wang, Y., Huang, S. Y. and Wu, C. M. 2010b. Characterization of antimicrobial resistance and molecular determinants of beta-lactamase in Escherichia coli isolated from chickens in China during 1970-2007. *Vet Microbiol.* 144, 505-510.

Li, L., Wang, B., Feng, S., Li, J., Wu, C., Wang, Y., Ruan, X. and Zeng, M. 2014. Prevalence and characteristics of extended-spectrum beta-lactamase and plasmid-mediated fluoroquinolone resistance genes in Escherichia coli isolated from chickens in Anhui province, China. *PLoS One.* 9, e104356.

Li, R., Chan, E. W. and Chen, S. 2016. Characterisation of a chromosomally-encoded extended-spectrum beta-lactamase gene blaPER-3 in Aeromonas caviae of chicken origin. *Int J Antimicrob Agents.* 47, 103-105.

Li, S., Song, W., Zhou, Y., Tang, Y., Gao, Y. and Miao, Z. 2015. Spread of extended-spectrum beta-lactamase-producing Escherichia coli from a swine farm to the receiving river. *Environ Sci Pollut Res Int.* 22, 13033-13037.

Liao, X. P., Liu, B. T., Yang, Q. E., Sun, J., Li, L., Fang, L. X. and Liu, Y. H. 2013. Comparison of plasmids coharboring 16s rrna methylase and extended-spectrum beta-lactamase genes among Escherichia coli isolates from pets and poultry. *J Food Prot.* 76, 2018-2023.

Liebana, E., Batchelor, M., Hopkins, K. L., Clifton-Hadley, F. A., Teale, C. J., Foster, A., Barker, L., Threlfall, E. J. and Davies, R. H. 2006. Longitudinal farm study of extended-spectrum beta-lactamase-mediated resistance. *J Clin Microbiol.* 44, 1630-1634.

Lim, E. J., Ho, S. X., Cao, D. Y., Lau, Q. C., Koh, T. H. and Hsu, L. Y. 2016. Extended-Spectrum Beta-Lactamase-Producing Enterobacteriaceae in Retail Chicken Meat in Singapore. *Ann Acad Med Singapore.* 45, 557-559.

Lim, J. S., Choi, D. S., Kim, Y. J., Chon, J. W., Kim, H. S., Park, H. J., Moon, J. S., Wee, S. H. and Seo, K. H. 2015. Characterization of Escherichia coli-Producing Extended-Spectrum beta-Lactamase (ESBL) Isolated from Chicken Slaughterhouses in South Korea. *Foodborne Pathog Dis.* 12, 741-748.

Lim, S. K., Lee, H. S., Nam, H. M., Jung, S. C. and Bae, Y. C. 2009. CTX-M-type beta-lactamase in Escherichia coli isolated from sick animals in Korea. *Microb Drug Resist.* 15, 139-142.

Lima Barbieri, N., Nielsen, D. W., Wannemuehler, Y., Cavender, T., Hussein, A., Yan, S. G., Nolan, L. K. and Logue, C. M. 2017. mcr-1 identified in Avian Pathogenic Escherichia coli (APEC). *PLoS One.* 12, e0172997.

Liu, J. H., Wei, S. Y., Ma, J. Y., Zeng, Z. L., Lu, D. H., Yang, G. X. and Chen, Z. L. 2007. Detection and characterisation of CTX-M and CMY-2 beta-lactamases among Escherichia coli isolates from farm animals in Guangdong Province of China. *Int J Antimicrob Agents.* 29, 576-581.

Locatelli, C., Caronte, I., Scaccabarozzi, L., Migliavacca, R., Pagani, L. and Moroni, P. 2009. Extended-spectrum beta-lactamase production in E. coli strains isolated from clinical bovine mastitis. *Vet Res Commun.* 33 Suppl 1, 141-144.

Locatelli, C., Scaccabarozzi, L., Pisoni, G. and Moroni, P. 2010. CTX-M1 ESBL-producing Klebsiella pneumoniae subsp. pneumoniae isolated from cases of bovine mastitis. *J Clin Microbiol.* 48, 3822-3823.

Lopez-Cerero, L., Egea, P., Serrano, L., Navarro, D., Mora, A., Blanco, J., Doi, Y., Paterson, D. L., Rodriguez-Bano, J. and Pascual, A. 2011. Characterisation of clinical and food animal Escherichia coli isolates producing CTX-M-15 extended-spectrum beta-lactamase belonging to ST410 phylogroup A. *Int J Antimicrob Agents.* 37, 365-367.

Lv, L., Partridge, S. R., He, L., Zeng, Z., He, D., Ye, J. and Liu, J. H. 2013. Genetic characterization of IncI2 plasmids carrying blaCTX-M-55 spreading in both pets and food animals in China. *Antimicrob Agents Chemother.* 57, 2824-2827.

Ma, J., Liu, J. H., Lv, L., Zong, Z., Sun, Y., Zheng, H., Chen, Z. and Zeng, Z. L. 2012. Characterization of extended-spectrum beta-lactamase genes found among Escherichia coli isolates from duck and environmental samples obtained on a duck farm. *Appl Environ Microbiol.* 78, 3668-3673.

Maamar, E., Hammami, S., Alonso, C. A., Dakhli, N., Abbassi, M. S., Ferjani, S., Hamzaoui, Z., Saidani, M., Torres, C. and Boutiba-Ben

Boubaker, I. 2016. High prevalence of extended-spectrum and plasmidic AmpC beta-lactamase-producing Escherichia coli from poultry in Tunisia. *Int J Food Microbiol.* 231, 69-75.

Machado, E., Coque, T. M., Canton, R., Sousa, J. C. and Peixe, L. 2008. Antibiotic resistance integrons and extended-spectrum {beta}-lactamases among Enterobacteriaceae isolates recovered from chickens and swine in Portugal. *J Antimicrob Chemother.* 62, 296-302.

Madec, J. Y., Doublet, B., Ponsin, C., Cloeckaert, A. and Haenni, M. 2011. Extended-spectrum beta-lactamase blaCTX-M-1 gene carried on an IncI1 plasmid in multidrug-resistant Salmonella enterica serovar Typhimurium DT104 in cattle in France. *J Antimicrob Chemother.* 66, 942-944.

Madec, J. Y., Haenni, M., Nordmann, P. and Poirel, L. 2017. Extended-spectrum beta-lactamase/AmpC- and carbapenemase-producing Enterobacteriaceae in animals: a threat for humans? *Clin Microbiol Infect.* 23, 826-833.

Madec, J. Y., Lazizzera, C., Chatre, P., Meunier, D., Martin, S., Lepage, G., Menard, M. F., Lebreton, P. and Rambaud, T. 2008. Prevalence of fecal carriage of acquired expanded-spectrum cephalosporin resistance in Enterobacteriaceae strains from cattle in France. *J Clin Microbiol.* 46, 1566-1567.

Mahanti, A., Ghosh, P., Samanta, I., Joardar, S. N., Bandyopadhyay, S., Bhattacharyya, D., Banerjee, J., Batabyal, S., Sar, T. K. and Dutta, T. K. 2017. Prevalence of CTX-M-Producing Klebsiella spp. in Broiler, Kuroiler, and Indigenous Poultry in West Bengal State, India. Microb Drug Resist.

Malhotra-Kumar, S., Xavier, B. B., Das, A. J., Lammens, C., Hoang, H. T., Pham, N. T. and Goossens, H. 2016. Colistin-resistant Escherichia coli harbouring mcr-1 isolated from food animals in Hanoi, Vietnam. *Lancet Infect Dis.* 16, 286-287.

Mamza, S. A., Egwu, G. O. and Mshelia, G. D. 2010. Beta-lactamase Escherichia coli and Staphylococcus aureus isolated from chickens in Nigeria. *Vet Ital.* 46, 155-165.

Matsumoto, Y., Izumiya, H., Sekizuka, T., Kuroda, M. and Ohnishi, M. 2014. Characterization of blaTEM-52-carrying plasmids of extended-spectrum-beta-lactamase-producing Salmonella enterica isolates from chicken meat with a common supplier in Japan. *Antimicrob Agents Chemother.* 58, 7545-7547.

Matsumoto, Y., Kitazume, H., Yamada, M., Ishiguro, Y., Muto, T., Izumiya, H. and Watanabe, H. 2007. CTX-M-14 type beta-lactamase producing Salmonella enterica serovar enteritidis isolated from imported chicken meat. *Jpn J Infect Dis.* 60, 236-238.

Mesa, R. J., Blanc, V., Blanch, A. R., Cortes, P., Gonzalez, J. J., Lavilla, S., Miro, E., Muniesa, M., Saco, M., Tortola, M. T., Mirelis, B., Coll, P., Llagostera, M., Prats, G. and Navarro, F. 2006. Extended-spectrum beta-lactamase-producing Enterobacteriaceae in different environments (humans, food, animal farms and sewage). *J Antimicrob Chemother.* 58, 211-215.

Meunier, D., Jouy, E., Lazizzera, C., Doublet, B., Kobisch, M., Cloeckaert, A. and Madec, J. Y. 2010. Plasmid-borne florfenicol and ceftiofur resistance encoded by the floR and blaCMY-2 genes in Escherichia coli isolates from diseased cattle in France. *J Med Microbiol.* 59, 467-471.

Meunier, D., Jouy, E., Lazizzera, C., Kobisch, M. and Madec, J. Y. 2006. CTX-M-1- and CTX-M-15-type beta-lactamases in clinical Escherichia coli isolates recovered from food-producing animals in France. *Int J Antimicrob Agents.* 28, 402-407.

Mevius, D. J., Koene, M. G. J., Wit, B., Pelt, W. V. and Bondt, N. 2011. *Monitoring of antimicrobial resistance and antibiotic usage in animals in the Netherlands in 2009.* Central Veterinary Institute of Wageningen University and Research Centre, Netherlands.

Meyer, E., Gastmeier, P., Kola, A. and Schwab, F. 2012. Pet animals and foreign travel are risk factors for colonisation with extended-spectrum beta-lactamase-producing Escherichia coli. *Infection.* 40, 685-687.

Mezhoud, H., Boyen, F., Touazi, L. H., Garmyn, A., Moula, N., Smet, A., Haesbrouck, F., Martel, A., Iguer-Ouada, M. and Touati, A. 2015. Extended spectrum beta-lactamase producing Escherichia coli in broiler breeding roosters: Presence in the reproductive tract and effect on sperm motility. *Anim Reprod Sci.* 159, 205-211.

Mezhoud, H., Chantziaras, I., Iguer-Ouada, M., Moula, N., Garmyn, A., Martel, A., Touati, A., Smet, A., Haesebrouck, F. and Boyen, F. 2016. Presence of antimicrobial resistance in coliform bacteria from hatching broiler eggs with emphasis on ESBL/AmpC-producing bacteria. *Avian Pathol.* 45, 493-500.

Michael, G. B., Kaspar, H., Siqueira, A. K., De Freitas Costa, E., Corbellini, L. G., Kadlec, K. and Schwarz, S. 2017. Extended-spectrum beta-lactamase (ESBL)-producing Escherichia coli isolates collected from diseased food-producing animals in the GERM-Vet monitoring program 2008-2014. *Vet Microbiol.* 200, 142-150.

Mnif, B., Ktari, S., Rhimi, F. M. and Hammami, A. 2012. Extensive dissemination of CTX-M-1- and CMY-2-producing Escherichia coli in poultry farms in Tunisia. *Lett Appl Microbiol.* 55, 407-413.

Molina-Lopez, J., Aparicio-Ozores, G., Ribas-Aparicio, R. M., Gavilanes-Parra, S., Chavez-Berrocal, M. E., Hernandez-Castro, R. and Manjarrez-Hernandez, H. A. 2011. Drug resistance, serotypes, and phylogenetic groups among uropathogenic Escherichia coli including O25-ST131 in Mexico City. *J Infect Dev Ctries.* 5, 840-849.

Mollenkopf, D. F., Cenera, J. K., Bryant, E. M., King, C. A., Kashoma, I., Kumar, A., Funk, J. A., Rajashekara, G. and Wittum, T. E. 2014. Organic or antibiotic-free labeling does not impact the recovery of enteric pathogens and antimicrobial-resistant Escherichia coli from fresh retail chicken. *Foodborne Pathog Dis.* 11, 920-929.

Mollenkopf, D. F., Weeman, M. F., Daniels, J. B., Abley, M. J., Mathews, J. L., Gebreyes, W. A. and Wittum, T. E. 2012. Variable within- and between-herd diversity of CTX-M cephalosporinase-bearing Escherichia coli isolates from dairy cattle. *Appl Environ Microbiol.* 78, 4552-4560.

Monte, D. F., Mem, A., Fernandes, M. R., Cerdeira, L., Esposito, F., Galvao, J. A., Franco, B., Lincopan, N. and Landgraf, M. 2017. Chicken Meat as a Reservoir of Colistin-Resistant Escherichia coli Strains Carrying mcr-1 Genes in South America. *Antimicrob Agents Chemother.* 61.

Moodley, A. and Guardabassi, L. 2009. Transmission of IncN plasmids carrying blaCTX-M-1 between commensal Escherichia coli in pigs and farm workers. *Antimicrob Agents Chemother.* 53, 1709-1711.

Mora, A., Blanco, M., Lopez, C., Mamani, R., Blanco, J. E., Alonso, M. P., Garcia-Garrote, F., Dahbi, G., Herrera, A., Fernandez, A., Fernandez, B., Agulla, A., Bou, G. and Blanco, J. 2011. Emergence of clonal groups O1:HNM-D-ST59, O15:H1-D-ST393, O20:H34/HNM-D-ST354, O25b:H4-B2-ST131 and ONT:H21,42-B1-ST101 among CTX-M-14-producing Escherichia coli clinical isolates in Galicia, northwest Spain. *Int J Antimicrob Agents.* 37, 16-21.

Mora, A., Herrera, A., Mamani, R., Lopez, C., Alonso, M. P., Blanco, J. E., Blanco, M., Dahbi, G., Garcia-Garrote, F., Pita, J. M., Coira, A., Bernardez, M. I. and Blanco, J. 2010. Recent emergence of clonal group O25b:K1:H4-B2-ST131 ibeA strains among Escherichia coli poultry isolates, including CTX-M-9-producing strains, and comparison with clinical human isolates. *Appl Environ Microbiol.* 76, 6991-6997.

Mulvey, M. R., Susky, E., Mccracken, M., Morck, D. W. and Read, R. R. 2009. Similar cefoxitin-resistance plasmids circulating in Escherichia coli from human and animal sources. *Vet Microbiol.* 134, 279-287.

Nakayama, T., Jinnai, M., Kawahara, R., Diep, K. T., Thang, N. N., Hoa, T. T., Hanh, L. K., Khai, P. N., Sumimura, Y. and Yamamoto, Y. 2017. Frequent use of colistin-based drug treatment to eliminate extended-spectrum beta-lactamase-producing Escherichia coli in backyard chicken farms in Thai Binh Province, Vietnam. *Trop Anim Health Prod.* 49, 31-37.

Nguyen Do, P., Nguyen, T. A., Le, T. H., Tran, N. M., Ngo, T. P., Dang, V. C., Kawai, T., Kanki, M., Kawahara, R., Jinnai, M., Yonogi, S., Hirai, Y., Yamamoto, Y. and Kumeda, Y. 2016. Dissemination of Extended-Spectrum beta-Lactamase- and AmpC beta-Lactamase-Producing Escherichia coli within the Food Distribution System of Ho Chi Minh City, Vietnam. *Biomed Res Int.* 2016, 8182096.

Norizuki, C., Kawamura, K., Wachino, J. I., Suzuki, M., Nagano, N., Kondo, T. and Arakawa, Y. 2017. Detection of Escherichia coli producing CTX-

M-1-group extended-spectrum beta-lactamases from pigs in Aichi prefecture, Japan, between 2015 and 2016. *Jpn J Infect Dis.*

Odenthal, S., Akineden, O. and Usleber, E. 2016. Extended-spectrum beta-lactamase producing Enterobacteriaceae in bulk tank milk from German dairy farms. *Int J Food Microbiol.* 238, 72-78.

Ohnishi, M., Okatani, A. T., Harada, K., Sawada, T., Marumo, K., Murakami, M., Sato, R., Esaki, H., Shimura, K., Kato, H., Uchida, N. and Takahashi, T. 2013. Genetic characteristics of CTX-M-type extended-spectrum-beta-lactamase (ESBL)-producing enterobacteriaceae involved in mastitis cases on Japanese dairy farms, 2007 to 2011. *J Clin Microbiol.* 51, 3117-3122.

Ojo, O. E., Schwarz, S. and Michael, G. B. 2016. Detection and characterization of extended-spectrum beta-lactamase-producing Escherichia coli from chicken production chains in Nigeria. *Vet Microbiol.* 194, 62-68.

Overdevest, I., Willemsen, I., Rijnsburger, M., Eustace, A., Xu, L., Hawkey, P., Heck, M., Savelkoul, P., Vandenbroucke-Grauls, C., Van Der Zwaluw, K., Huijsdens, X. and Kluytmans, J. 2011. Extended-spectrum beta-lactamase genes of Escherichia coli in chicken meat and humans, The Netherlands. *Emerg Infect Dis.* 17, 1216-1222.

Pacholewicz, E., Liakopoulos, A., Swart, A., Gortemaker, B., Dierikx, C., Havelaar, A. and Schmitt, H. 2015. Reduction of extended-spectrum-beta-lactamase- and AmpC-beta-lactamase-producing Escherichia coli through processing in two broiler chicken slaughterhouses. *Int J Food Microbiol.* 215, 57-63.

Paivarinta, M., Pohjola, L., Fredriksson-Ahomaa, M. and Heikinheimo, A. 2016. Low Occurrence of Extended-Spectrum beta-lactamase-Producing Escherichia coli in Finnish Food-Producing Animals. *Zoonoses Public Health.* 63, 624-631.

Park, Y. S., Adams-Haduch, J. M., Rivera, J. I., Curry, S. R., Harrison, L. H. and Doi, Y. 2012. Escherichia coli producing CMY-2 beta-lactamase in retail chicken, Pittsburgh, Pennsylvania, USA. *Emerg Infect Dis.* 18, 515-516.

Park, Y. S., Yoo, S., Seo, M. R., Kim, J. Y., Cho, Y. K. and Pai, H. 2009. Risk factors and clinical features of infections caused by plasmid-mediated AmpC beta-lactamase-producing Enterobacteriaceae. *Int J Antimicrob Agents.* 34, 38-43.

Pehlivanoglu, F., Turutoglu, H. and Ozturk, D. 2016. CTX-M-15-Type Extended-Spectrum Beta-Lactamase-Producing Escherichia coli as Causative Agent of Bovine Mastitis. *Foodborne Pathog Dis.* 13, 477-482.

Persoons, D., Haesebrouck, F., Smet, A., Herman, L., Heyndrickx, M., Martel, A., Catry, B., Berge, A. C., Butaye, P. and Dewulf, J. 2011. Risk factors for ceftiofur resistance in Escherichia coli from Belgian broilers. *Epidemiol Infect.* 139, 765-771.

Pitout, J. D. 2013. Enterobacteriaceae that produce extended-spectrum beta-lactamases and AmpC beta-lactamases in the community: the tip of the iceberg? *Curr Pharm Des.* 19, 257-263.

Pitout, J. D. and Laupland, K. B. 2008. Extended-spectrum beta-lactamase-producing Enterobacteriaceae: an emerging public-health concern. *Lancet Infect Dis.* 8, 159-166.

Platell, J. L., Johnson, J. R., Cobbold, R. N. and Trott, D. J. 2011. Multidrug-resistant extraintestinal pathogenic Escherichia coli of sequence type ST131 in animals and foods. *Vet Microbiol.* 153, 99-108.

Politi, L., Tassios, P. T., Lambiri, M., Kansouzidou, A., Pasiotou, M., Vatopoulos, A. C., Mellou, K., Legakis, N. J. and Tzouvelekis, L. S. 2005. Repeated occurrence of diverse extended-spectrum beta-lactamases in minor serotypes of food-borne Salmonella enterica subsp. enterica. *J Clin Microbiol.* 43, 3453-3456.

Poole, T. L., Edrington, T. S., Brichta-Harhay, D. M., Carattoli, A., Anderson, R. C. and Nisbet, D. J. 2009. Conjugative transferability of the A/C plasmids from Salmonella enterica isolates that possess or lack bla(CMY) in the A/C plasmid backbone. *Foodborne Pathog Dis.* 6, 1185-1194.

Projahn, M., Daehre, K., Roesler, U. and Friese, A. 2017. Extended-Spectrum-Beta-Lactamase- and Plasmid-Encoded Cephamycinase-

Producing Enterobacteria in the Broiler Hatchery as a Potential Mode of Pseudo-Vertical Transmission. *Appl Environ Microbiol.* 83.

Ramos, S., Silva, N., Dias, D., Sousa, M., Capelo-Martinez, J. L., Brito, F., Canica, M., Igrejas, G. and Poeta, P. 2013. Clonal diversity of ESBL-producing Escherichia coli in pigs at slaughter level in Portugal. *Foodborne Pathog Dis.* 10, 74-79.

Randall, L., Heinrich, K., Horton, R., Brunton, L., Sharman, M., Bailey-Horne, V., Sharma, M., Mclaren, I., Coldham, N., Teale, C. and Jones, J. 2014a. Detection of antibiotic residues and association of cefquinome residues with the occurrence of Extended-Spectrum beta-Lactamase (ESBL)-producing bacteria in waste milk samples from dairy farms in England and Wales in 2011. *Res Vet Sci.* 96, 15-24.

Randall, L. P., Clouting, C., Horton, R. A., Coldham, N. G., Wu, G., Clifton-Hadley, F. A., Davies, R. H. and Teale, C. J. 2011. Prevalence of Escherichia coli carrying extended-spectrum beta-lactamases (CTX-M and TEM-52) from broiler chickens and turkeys in Great Britain between 2006 and 2009. *J Antimicrob Chemother.* 66, 86-95.

Randall, L. P., Lemma, F., Rogers, J. P., Cheney, T. E., Powell, L. F. and Teale, C. J. 2014b. Prevalence of extended-spectrum-beta-lactamase-producing Escherichia coli from pigs at slaughter in the UK in 2013. *J Antimicrob Chemother.* 69, 2947-2950.

Rankin, S. C., Aceto, H., Cassidy, J., Holt, J., Young, S., Love, B., Tewari, D., Munro, D. S. and Benson, C. E. 2002. Molecular characterization of cephalosporin-resistant Salmonella enterica serotype Newport isolates from animals in Pennsylvania. *J Clin Microbiol.* 40, 4679-4684.

Rasmussen, M. M., Opintan, J. A., Frimodt-Moller, N. and Styrishave, B. 2015. Beta-Lactamase Producing Escherichia coli Isolates in Imported and Locally Produced Chicken Meat from Ghana. PLoS One. 10, e0139706.

Rayamajhi, N., Jung, B. Y., Cha, S. B., Shin, M. K., Kim, A., Kang, M. S., Lee, K. M. and Yoo, H. S. 2010. Antibiotic resistance patterns and detection of blaDHA-1 in Salmonella species isolates from chicken farms in South Korea. *Appl Environ Microbiol.* 76, 4760-4764.

Rayamajhi, N., Kang, S. G., Lee, D. Y., Kang, M. L., Lee, S. I., Park, K. Y., Lee, H. S. and Yoo, H. S. 2008. Characterization of TEM-, SHV- and AmpC-type beta-lactamases from cephalosporin-resistant Enterobacteriaceae isolated from swine. *Int J Food Microbiol.* 124, 183-187.

Reddy, P., Malczynski, M., Obias, A., Reiner, S., Jin, N., Huang, J., Noskin, G. A. and Zembower, T. 2007. Screening for extended-spectrum beta-lactamase-producing Enterobacteriaceae among high-risk patients and rates of subsequent bacteremia. *Clin Infect Dis.* 45, 846-852.

Reich, F., Atanassova, V. and Klein, G. 2013. Extended-spectrum beta-lactamase- and AmpC-producing enterobacteria in healthy broiler chickens, Germany. *Emerg Infect Dis.* 19, 1253-1259.

Reist, M., Geser, N., Hachler, H., Scharrer, S. and Stephan, R. 2013. ESBL-producing Enterobacteriaceae: occurrence, risk factors for fecal carriage and strain traits in the Swiss slaughter cattle population younger than 2 years sampled at abattoir level. *PLoS One.* 8, e71725.

Riano, I., Moreno, M. A., Teshager, T., Saenz, Y., Dominguez, L. and Torres, C. 2006. Detection and characterization of extended-spectrum beta-lactamases in Salmonella enterica strains of healthy food animals in Spain. *J Antimicrob Chemother.* 58, 844-847.

Robicsek, A., Strahilevitz, J., Jacoby, G. A., Macielag, M., Abbanat, D., Park, C. H., Bush, K. and Hooper, D. C. 2006. Fluoroquinolone-modifying enzyme: a new adaptation of a common aminoglycoside acetyltransferase. *Nat Med.* 12, 83-88.

Rodrigues, C., Machado, E., Peixe, L. and Novais, A. 2013. IncI1/ST3 and IncN/ST1 plasmids drive the spread of blaTEM-52 and blaCTX-M-1/-32 in diverse Escherichia coli clones from different piggeries. *J Antimicrob Chemother.* 68, 2245-2248.

Rodriguez, I., Barownick, W., Helmuth, R., Mendoza, M. C., Rodicio, M. R., Schroeter, A. and Guerra, B. 2009. Extended-spectrum {beta}-lactamases and AmpC {beta}-lactamases in ceftiofur-resistant Salmonella enterica isolates from food and livestock obtained in Germany during 2003-07. *J Antimicrob Chemother.* 64, 301-309.

Rubin, J. E. and Pitout, J. D. 2014. Extended-spectrum beta-lactamase, carbapenemase and AmpC producing Enterobacteriaceae in companion animals. *Vet Microbiol.* 170, 10-18.

Saishu, N., Ozaki, H. and Murase, T. 2014. CTX-M-type extended-spectrum beta-lactamase-producing Klebsiella pneumoniae isolated from cases of bovine mastitis in Japan. *J Vet Med Sci.* 76, 1153-1156.

Saliu, E. M., Vahjen, W. and Zentek, J. 2017. Types and prevalence of extended-spectrum beta-lactamase producing Enterobacteriaceae in poultry. *Anim Health Res Rev.* 18, 46-57.

Samanta, I., Joardar, S. N., Das, P. K. and Sar, T. K. 2015a. Comparative possession of Shiga toxin, intimin, enterohaemolysin and major extended spectrum beta lactamase (ESBL) genes in Escherichia coli isolated from backyard and farmed poultry. *Iran J Vet Res.* 16, 90-93.

Samanta, I., Joardar, S. N., Mahanti, A., Bandyopadhyay, S., Sar, T. K. and Dutta, T. K. 2015b. Approaches to characterize extended spectrum beta-lactamase/beta-lactamase producing Escherichia coli in healthy organized vis-a-vis backyard farmed pigs in India. *Infect Genet Evol.* 36, 224-230.

Santman-Berends, I. M., Gonggrijp, M. A., Hage, J. J., Heuvelink, A. E., Velthuis, A., Lam, T. J. and Van Schaik, G. 2017. Prevalence and risk factors for extended-spectrum beta-lactamase or AmpC-producing Escherichia coli in organic dairy herds in the Netherlands. *J Dairy Sci.* 100, 562-571.

Schaumburg, F., Alabi, A. S., Frielinghaus, L., Grobusch, M. P., Kock, R., Becker, K., Issifou, S., Kremsner, P. G., Peters, G. and Mellmann, A. 2014. The risk to import ESBL-producing Enterobacteriaceae and Staphylococcus aureus through chicken meat trade in Gabon. *BMC Microbiol.* 14, 286.

Schill, F., Abdulmawjood, A., Klein, G. and Reich, F. 2017. Prevalence and characterization of extended-spectrum beta-lactamase (ESBL) and AmpC beta-lactamase producing Enterobacteriaceae in fresh pork meat at processing level in Germany. *Int J Food Microbiol.* 257, 58-66.

Schink, A. K., Kadlec, K. and Schwarz, S. 2011. Analysis of bla(CTX-M)-carrying plasmids from Escherichia coli isolates collected in the BfT-GermVet study. *Appl Environ Microbiol.* 77, 7142-7146.

Schmid, A., Hormansdorfer, S., Messelhausser, U., Kasbohrer, A., Sauter-Louis, C. and Mansfeld, R. 2013. Prevalence of extended-spectrum beta-lactamase-producing Escherichia coli on Bavarian dairy and beef cattle farms. *Appl Environ Microbiol.* 79, 3027-3032.

Schmithausen, R. M., Schulze-Geisthoevel, S. V., Stemmer, F., El-Jade, M., Reif, M., Hack, S., Meilaender, A., Montabauer, G., Fimmers, R., Parcina, M., Hoerauf, A., Exner, M., Petersen, B., Bierbaum, G. and Bekeredjian-Ding, I. 2015. Analysis of Transmission of MRSA and ESBL-E among Pigs and Farm Personnel. *PLoS One.* 10, e0138173.

Schwaiger, K., Bauer, J. and Holzel, C. S. 2013. Selection and persistence of antimicrobial-resistant Escherichia coli including extended-spectrum beta-lactamase producers in different poultry flocks on one chicken farm. *Microb Drug Resist.* 19, 498-506.

Shahada, F., Chuma, T., Dahshan, H., Akiba, M., Sueyoshi, M. and Okamoto, K. 2010. Detection and characterization of extended-spectrum beta-lactamase (TEM-52)-producing Salmonella serotype Infantis from broilers in Japan. *Foodborne Pathog Dis.* 7, 515-521.

Sheikh, A. A., Checkley, S., Avery, B., Chalmers, G., Bohaychuk, V., Boerlin, P., Reid-Smith, R. and Aslam, M. 2012. Antimicrobial resistance and resistance genes in Escherichia coli isolated from retail meat purchased in Alberta, Canada. *Foodborne Pathog Dis.* 9, 625-631.

Shin, S. W., Jung, M., Shin, M. K. and Yoo, H. S. 2015. Profiling of antimicrobial resistance and plasmid replicon types in beta-lactamase producing Escherichia coli isolated from Korean beef cattle. *J Vet Sci.* 16, 483-489.

Shin, S. W., Jung, M., Won, H. G., Belaynehe, K. M., Yoon, I. J. and Yoo, H. S. 2017. Characteristics of Transmissible CTX-M- and CMY-Type beta-Lactamase-Producing Escherichia coli Isolates Collected from Pig and Chicken Farms in South Korea. *J Microbiol Biotechnol.* 27, 1716-1723.

Shiraki, Y., Shibata, N., Doi, Y. and Arakawa, Y. 2004. Escherichia coli producing CTX-M-2 beta-lactamase in cattle, Japan. *Emerg Infect Dis.* 10, 69-75.

Silva, K. C., Fontes, L. C., Moreno, A. M., Astolfi-Ferreira, C. S., Ferreira, A. J. and Lincopan, N. 2013. Emergence of extended-spectrum-beta-lactamase CTX-M-2-producing Salmonella enterica serovars Schwarzengrund and Agona in poultry farms. *Antimicrob Agents Chemother.* 57, 3458-3459.

Simoes, R. R., Poirel, L., Da Costa, P. M. and Nordmann, P. 2010. Seagulls and beaches as reservoirs for multidrug-resistant Escherichia coli. *Emerg Infect Dis.* 16, 110-112.

Smet, A., Martel, A., Persoons, D., Dewulf, J., Heyndrickx, M., Catry, B., Herman, L., Haesebrouck, F. and Butaye, P. 2008. Diversity of extended-spectrum beta-lactamases and class C beta-lactamases among cloacal Escherichia coli Isolates in Belgian broiler farms. *Antimicrob Agents Chemother.* 52, 1238-1243.

Smet, A., Martel, A., Persoons, D., Dewulf, J., Heyndrickx, M., Cloeckaert, A., Praud, K., Claeys, G., Catry, B., Herman, L., Haesebrouck, F. and Butaye, P. 2009. Comparative analysis of extended-spectrum-{beta}-lactamase-carrying plasmids from different members of Enterobacteriaceae isolated from poultry, pigs and humans: evidence for a shared {beta}-lactam resistance gene pool? *J Antimicrob Chemother.* 63, 1286-1288.

Smet, A., Rasschaert, G., Martel, A., Persoons, D., Dewulf, J., Butaye, P., Catry, B., Haesebrouck, F., Herman, L. and Heyndrickx, M. 2011. In situ ESBL conjugation from avian to human Escherichia coli during cefotaxime administration. *J Appl Microbiol.* 110, 541-549.

Snow, L. C., Warner, R. G., Cheney, T., Wearing, H., Stokes, M., Harris, K., Teale, C. J. and Coldham, N. G. 2012. Risk factors associated with extended spectrum beta-lactamase Escherichia coli (CTX-M) on dairy farms in North West England and North Wales. *Prev Vet Med.* 106, 225-234.

Snow, L. C., Wearing, H., Stephenson, B., Teale, C. J. and Coldham, N. G. 2011. Investigation of the presence of ESBL-producing Escherichia coli in the North Wales and West Midlands areas of the UK in 2007 to 2008 using scanning surveillance. *Vet Rec.* 169, 656.

Sola-Gines, M., Gonzalez-Lopez, J. J., Cameron-Veas, K., Piedra-Carrasco, N., Cerda-Cuellar, M. and Migura-Garcia, L. 2015. Houseflies (Musca domestica) as Vectors for Extended-Spectrum beta-Lactamase-Producing Escherichia coli on Spanish Broiler Farms. *Appl Environ Microbiol.* 81, 3604-3611.

Stromdahl, H., Tham, J., Melander, E., Walder, M., Edquist, P. J. and Odenholt, I. 2011. Prevalence of faecal ESBL carriage in the community and in a hospital setting in a county of Southern Sweden. *Eur J Clin Microbiol Infect Dis.* 30, 1159-1162.

Sudarwanto, M., Akineden, O., Odenthal, S., Gross, M. and Usleber, E. 2015. Extended-Spectrum beta-Lactamase (ESBL)-Producing Klebsiella pneumoniae in Bulk Tank Milk from Dairy Farms in Indonesia. *Foodborne Pathog Dis.* 12, 585-590.

Sugawara, M., Komori, J., Kawakami, M., Izumiya, H., Watanabe, H. and Akiba, M. 2011. Molecular and phenotypic characteristics of CMY-2 beta-lactamase-producing Salmonella enterica serovar Typhimurium isolated from cattle in Japan. *J Vet Med Sci.* 73, 345-349.

SVARM. 2011. Swedish veterinary antimicrobial resistance monitoring. SVARM, Uppsala, Sweden.

SWEDRES-SVARM. 2014. Consumption of antibiotics and occurrence of antibiotic resistance in Sweden. Swedres-Svarm.

Taguchi, M., Kawahara, R., Seto, K., Harada, T. and Kumeda, Y. 2012. Extended-spectrum beta-lactamase- and AmpC beta-lactamase-producing Salmonella enterica strains isolated from domestic retail chicken meat from 2006 to 2011. *Jpn J Infect Dis.* 65, 555-557.

Taguchi, M., Seto, K., Yamazaki, W., Tsukamoto, T., Izumiya, H. and Watanabe, H. 2006. CMY-2 beta-lactamase-producing Salmonella enterica serovar infantis isolated from poultry in Japan. *Jpn J Infect Dis.* 59, 144-146.

Tamang, M. D., Nam, H. M., Gurung, M., Jang, G. C., Kim, S. R., Jung, S. C., Park, Y. H. and Lim, S. K. 2013a. Molecular characterization of CTX-M beta-lactamase and associated addiction systems in Escherichia coli circulating among cattle, farm workers, and the farm environment. *Appl Environ Microbiol.* 79, 3898-3905.

Tamang, M. D., Nam, H. M., Kim, S. R., Chae, M. H., Jang, G. C., Jung, S. C. and Lim, S. K. 2013b. Prevalence and molecular characterization of CTX-M beta-lactamase-producing Escherichia coli isolated from healthy swine and cattle. *Foodborne Pathog Dis.* 10, 13-20.

Tamang, M. D., Nam, H. M., Kim, T. S., Jang, G. C., Jung, S. C. and Lim, S. K. 2011. Emergence of extended-spectrum beta-lactamase (CTX-M-15 and CTX-M-14)-producing nontyphoid Salmonella with reduced susceptibility to ciprofloxacin among food animals and humans in Korea. *J Clin Microbiol.* 49, 2671-2675.

Tangden, T., Cars, O., Melhus, A. and Lowdin, E. 2010. Foreign travel is a major risk factor for colonization with Escherichia coli producing CTX-M-type extended-spectrum beta-lactamases: a prospective study with Swedish volunteers. *Antimicrob Agents Chemother.* 54, 3564-3568.

Tate, H., Folster, J. P., Hsu, C. H., Chen, J., Hoffmann, M., Li, C., Morales, C., Tyson, G. H., Mukherjee, S., Brown, A. C., Green, A., Wilson, W., Dessai, U., Abbott, J., Joseph, L., Haro, J., Ayers, S., Mcdermott, P. F. and Zhao, S. 2017. Comparative Analysis of Extended-Spectrum-beta-Lactamase CTX-M-65-Producing Salmonella enterica Serovar Infantis Isolates from Humans, Food Animals, and Retail Chickens in the United States. *Antimicrob Agents Chemother.* 61.

Thomson, K. S. 2010. Extended-spectrum-beta-lactamase, AmpC, and Carbapenemase issues. *J Clin Microbiol.* 48, 1019-1025.

Tian, G. B., Wang, H. N., Zhang, A. Y., Zhang, Y., Fan, W. Q., Xu, C. W., Zeng, B., Guan, Z. B. and Zou, L. K. 2012. Detection of clinically important beta-lactamases in commensal Escherichia coli of human and swine origin in western China. *J Med Microbiol.* 61, 233-238.

Tian, G. B., Wang, H. N., Zou, L. K., Tang, J. N., Zhao, Y. W., Ye, M. Y., Tang, J. Y., Zhang, Y., Zhang, A. Y., Yang, X., Xu, C. W. and Fu, Y. J. 2009. Detection of CTX-M-15, CTX-M-22, and SHV-2 extended-spectrum beta-lactamases (ESBLs) in Escherichia coli fecal-sample isolates from pig farms in China. *Foodborne Pathog Dis.* 6, 297-304.

Tong, P., Sun, Y., Ji, X., Du, X., Guo, X., Liu, J., Zhu, L., Zhou, B., Zhou, W., Liu, G. and Feng, S. 2015. Characterization of antimicrobial resistance and extended-spectrum beta-lactamase genes in Escherichia coli isolated from chickens. *Foodborne Pathog Dis.* 12, 345-352.

Tseng, S. P., Wang, S. F., Kuo, C. Y., Huang, J. W., Hung, W. C., Ke, G. M. and Lu, P. L. 2015. Characterization of Fosfomycin Resistant Extended-Spectrum beta-Lactamase-Producing Escherichia coli Isolates from Human and Pig in Taiwan. *PLoS One.* 10, e0135864.

Usui, M., Iwasa, T., Fukuda, A., Sato, T., Okubo, T. and Tamura, Y. 2013. The role of flies in spreading the extended-spectrum beta-lactamase gene from cattle. *Microb Drug Resist.* 19, 415-420.

Valat, C., Auvray, F., Forest, K., Metayer, V., Gay, E., Peytavin De Garam, C., Madec, J. Y. and Haenni, M. 2012. Phylogenetic grouping and virulence potential of extended-spectrum-beta-lactamase-producing Escherichia coli strains in cattle. *Appl Environ Microbiol.* 78, 4677-4682.

Valverde, A., Coque, T. M., Sanchez-Moreno, M. P., Rollan, A., Baquero, F. and Canton, R. 2004. Dramatic increase in prevalence of fecal carriage of extended-spectrum beta-lactamase-producing Enterobacteriaceae during nonoutbreak situations in Spain. *J Clin Microbiol.* 42, 4769-4775.

Van Damme, I., Garcia-Graells, C., Biasino, W., Gowda, T., Botteldoorn, N. and De Zutter, L. 2017. High abundance and diversity of extended-spectrum beta-lactamase (ESBL)-producing Escherichia coli in faeces and tonsils of pigs at slaughter. *Vet Microbiol.* 208, 190-194.

Van, T. T., Nguyen, H. N., Smooker, P. M. and Coloe, P. J. 2012. The antibiotic resistance characteristics of non-typhoidal Salmonella enterica isolated from food-producing animals, retail meat and humans in South East Asia. *Int J Food Microbiol.* 154, 98-106.

Vincent, C., Boerlin, P., Daignault, D., Dozois, C. M., Dutil, L., Galanakis, C., Reid-Smith, R. J., Tellier, P. P., Tellis, P. A., Ziebell, K. and Manges, A. R. 2010. Food reservoir for Escherichia coli causing urinary tract infections. *Emerg Infect Dis.* 16, 88-95.

Voets, G. M., Fluit, A. C., Scharringa, J., Schapendonk, C., Van Den Munckhof, T., Leverstein-Van Hall, M. A. and Stuart, J. C. 2013. Identical plasmid AmpC beta-lactamase genes and plasmid types in E. coli isolates from patients and poultry meat in the Netherlands. *Int J Food Microbiol.* 167, 359-362.

Vogt, D., Overesch, G., Endimiani, A., Collaud, A., Thomann, A. and Perreten, V. 2014. Occurrence and genetic characteristics of third-generation cephalosporin-resistant Escherichia coli in Swiss retail meat. *Microb Drug Resist.* 20, 485-494.

Von Salviati, C., Friese, A., Roschanski, N., Laube, H., Guerra, B., Kasbohrer, A., Kreienbrock, L. and Roesler, U. 2014. Extended-spectrum beta-lactamases (ESBL)/AmpC beta-lactamases-producing Escherichia coli in German fattening pig farms: a longitudinal study. *Berl Munch Tierarztl Wochenschr.* 127, 412-419.

Von Salviati, C., Laube, H., Guerra, B., Roesler, U. and Friese, A. 2015. Emission of ESBL/AmpC-producing Escherichia coli from pig fattening farms to surrounding areas. *Vet Microbiol.* 175, 77-84.

Warren, R. E., Ensor, V. M., O'neill, P., Butler, V., Taylor, J., Nye, K., Harvey, M., Livermore, D. M., Woodford, N. and Hawkey, P. M. 2008. Imported chicken meat as a potential source of quinolone-resistant Escherichia coli producing extended-spectrum beta-lactamases in the UK. *J Antimicrob Chemother.* 61, 504-508.

Wasiński, B., Różańska, H. and Osek, J. 2013. Occurrence of extended spectrum ß-Lactamaseand AmpC-Producing Escherichia coli in meat samples. *Bulletin- Veterinary Institute in Pulawy.* 57, 513-517.

Wasyl, D., Hasman, H., Cavaco, L. M. and Aarestrup, F. M. 2012. Prevalence and characterization of cephalosporin resistance in nonpathogenic Escherichia coli from food-producing animals slaughtered in Poland. *Microb Drug Resist.* 18, 79-82.

Wasyl, D. and Hoszowski, A. 2012. First isolation of ESBL-producing Salmonella and emergence of multiresistant Salmonella Kentucky in turkey in Poland. *Food Research International.* 45, 958-961.

Watson, E., Jeckel, S., Snow, L., Stubbs, R., Teale, C., Wearing, H., Horton, R., Toszeghy, M., Tearne, O., Ellis-Iversen, J. and Coldham, N. 2012.

Epidemiology of extended spectrum beta-lactamase E. coli (CTX-M-15) on a commercial dairy farm. *Vet Microbiol.* 154, 339-346.

Weill, F. X., Lailler, R., Praud, K., Kerouanton, A., Fabre, L., Brisabois, A., Grimont, P. A. and Cloeckaert, A. 2004. Emergence of extended-spectrum-beta-lactamase (CTX-M-9)-producing multiresistant strains of Salmonella enterica serotype Virchow in poultry and humans in France. *J Clin Microbiol.* 42, 5767-5773.

Wener, K. M., Schechner, V., Gold, H. S., Wright, S. B. and Carmeli, Y. 2010. Treatment with fluoroquinolones or with beta-lactam-beta-lactamase inhibitor combinations is a risk factor for isolation of extended-spectrum-beta-lactamase-producing Klebsiella species in hospitalized patients. *Antimicrob Agents Chemother.* 54, 2010-2016.

White, D. G., Hudson, C., Maurer, J. J., Ayers, S., Zhao, S., Lee, M. D., Bolton, L., Foley, T. and Sherwood, J. 2000. Characterization of chloramphenicol and florfenicol resistance in Escherichia coli associated with bovine diarrhea. *J Clin Microbiol.* 38, 4593-4598.

Wieler, L. H., Semmler, T., Eichhorn, I., Antao, E. M., Kinnemann, B., Geue, L., Karch, H., Guenther, S. and Bethe, A. 2011. No evidence of the Shiga toxin-producing E. coli O104:H4 outbreak strain or enteroaggregative E. coli (EAEC) found in cattle faeces in northern Germany, the hotspot of the 2011 HUS outbreak area. *Gut Pathog.* 3, 17.

Winokur, P. L., Brueggemann, A., Desalvo, D. L., Hoffmann, L., Apley, M. D., Uhlenhopp, E. K., Pfaller, M. A. and Doern, G. V. 2000. Animal and human multidrug-resistant, cephalosporin-resistant salmonella isolates expressing a plasmid-mediated CMY-2 AmpC beta-lactamase. *Antimicrob Agents Chemother.* 44, 2777-2783.

Winokur, P. L., Vonstein, D. L., Hoffman, L. J., Uhlenhopp, E. K. and Doern, G. V. 2001. Evidence for transfer of CMY-2 AmpC beta-lactamase plasmids between Escherichia coli and Salmonella isolates from food animals and humans. *Antimicrob Agents Chemother.* 45, 2716-2722.

Wittum, T. E., Mollenkopf, D. F., Daniels, J. B., Parkinson, A. E., Mathews, J. L., Fry, P. R., Abley, M. J. and Gebreyes, W. A. 2010. CTX-M-type extended-spectrum beta-lactamases present in Escherichia coli from the feces of cattle in Ohio, United States. *Foodborne Pathog Dis.* 7, 1575-1579.

Wu, H., Liu, B. G., Liu, J. H., Pan, Y. S., Yuan, L. and Hu, G. Z. 2012. Phenotypic and molecular characterization of CTX-M-14 extended-spectrum beta-lactamase and plasmid-mediated ACT-like AmpC beta-lactamase produced by Klebsiella pneumoniae isolates from chickens in Henan Province, China. *Genet Mol Res.* 11, 3357-3364.

Wu, H., Xia, X., Cui, Y., Hu, Y., Xi, M., Wang, X., Shi, X., Wang, D., Meng, J. and Yang, B. 2013. Prevalence of extended-spectrum b-lactamase-producing Salmonella on retail chicken in six provinces and two national cities in the People's Republic of China. *J Food Prot.* 76, 2040-2044.

Wu, S., Chouliara, E., Hasman, H., Dalsgaard, A., Vieira, A. and Jensen, L. B. 2008. Detection of a single isolate of CTX-M-1-producing Escherichia coli from healthy pigs in Denmark. *J Antimicrob Chemother.* 61, 747-749.

Yan, J. J., Hong, C. Y., Ko, W. C., Chen, Y. J., Tsai, S. H., Chuang, C. L. and Wu, J. J. 2004. Dissemination of blaCMY-2 among Escherichia coli isolates from food animals, retail ground meats, and humans in southern Taiwan. *Antimicrob Agents Chemother.* 48, 1353-1356.

Yang, B., Cui, Y., Shi, C., Wang, J., Xia, X., Xi, M., Wang, X., Meng, J., Alali, W. Q., Walls, I. and Doyle, M. P. 2014. Counts, serotypes, and antimicrobial resistance of Salmonella isolates on retail raw poultry in the People's Republic of China. *J Food Prot.* 77, 894-902.

Yoon, R. H., Cha, S. Y., Wei, B., Roh, J. H., Seo, H. S., Oh, J. Y. and Jang, H. K. 2014. Prevalence of Salmonella isolates and antimicrobial resistance in poultry meat from South Korea. *J Food Prot.* 77, 1579-1582.

Yuan, L., Liu, J. H., Hu, G. Z., Pan, Y. S., Liu, Z. M., Mo, J. and Wei, Y. J. 2009. Molecular characterization of extended-spectrum beta-lactamase-producing Escherichia coli isolates from chickens in Henan Province, China. *J Med Microbiol.* 58, 1449-1453.

Zaidi, M. B., Leon, V., Canche, C., Perez, C., Zhao, S., Hubert, S. K., Abbott, J., Blickenstaff, K. and Mcdermott, P. F. 2007. Rapid and widespread dissemination of multidrug-resistant blaCMY-2 Salmonella Typhimurium in Mexico. *J Antimicrob Chemother.* 60, 398-401.

Zhang, H., Zhai, Z., Li, Q., Liu, L., Guo, S., Li, Q., Yang, L., Ye, C., Chang, W. and Zhai, J. 2016. Characterization of Extended-Spectrum beta-Lactamase-Producing Escherichia coli Isolates from Pigs and Farm Workers. *J Food Prot.* 79, 1630-1634.

Zhang, W. J., Wang, X. M., Dai, L., Hua, X., Dong, Z., Schwarz, S. and Liu, S. 2015. Novel conjugative plasmid from Escherichia coli of swine origin that coharbors the multiresistance gene cfr and the extended-spectrum-beta-lactamase gene blaCTX-M-14b. *Antimicrob Agents Chemother.* 59, 1337-1340.

Zhao, S., Blickenstaff, K., Bodeis-Jones, S., Gaines, S. A., Tong, E. and Mcdermott, P. F. 2012. Comparison of the prevalences and antimicrobial resistances of Escherichia coli isolates from different retail meats in the United States, 2002 to 2008. *Appl Environ Microbiol.* 78, 1701-1707.

Zhao, S., Blickenstaff, K., Glenn, A., Ayers, S. L., Friedman, S. L., Abbott, J. W. and Mcdermott, P. F. 2009. beta-Lactam resistance in salmonella strains isolated from retail meats in the United States by the National Antimicrobial Resistance Monitoring System between 2002 and 2006. *Appl Environ Microbiol.* 75, 7624-7630.

Zhao, S., Qaiyumi, S., Friedman, S., Singh, R., Foley, S. L., White, D. G., Mcdermott, P. F., Donkar, T., Bolin, C., Munro, S., Baron, E. J. and Walker, R. D. 2003. Characterization of Salmonella enterica serotype newport isolated from humans and food animals. *J Clin Microbiol.* 41, 5366-5371.

Zhao, S., White, D. G., Friedman, S. L., Glenn, A., Blickenstaff, K., Ayers, S. L., Abbott, J. W., Hall-Robinson, E. and Mcdermott, P. F. 2008. Antimicrobial resistance in Salmonella enterica serovar Heidelberg isolates from retail meats, including poultry, from 2002 to 2006. *Appl Environ Microbiol.* 74, 6656-6662.

Zhao, S., White, D. G., Mcdermott, P. F., Friedman, S., English, L., Ayers, S., Meng, J., Maurer, J. J., Holland, R. and Walker, R. D. 2001. Identification and expression of cephamycinase bla(CMY) genes in

Escherichia coli and Salmonella isolates from food animals and ground meat. *Antimicrob Agents Chemother.* 45, 3647-3650.

Zheng, H., Zeng, Z., Chen, S., Liu, Y., Yao, Q., Deng, Y., Chen, X., Lv, L., Zhuo, C., Chen, Z. and Liu, J. H. 2012. Prevalence and characterisation of CTX-M beta-lactamases amongst Escherichia coli isolates from healthy food animals in China. *Int J Antimicrob Agents.* 39, 305-310.

Ziech, R. E., Lampugnani, C., Perin, A. P., Sereno, M. J., Sfaciotte, R. A., Viana, C., Soares, V. M., Pinto, J. P. and Bersot Ldos, S. 2016. Multidrug resistance and ESBL-producing Salmonella spp. isolated from broiler processing plants. *Braz J Microbiol.* 47, 191-195.

Zogg, A. L., Zurfluh, K., Nuesch-Inderbinen, M. and Stephan, R. 2016. Characteristics of ESBL-producing Enterobacteriaceae and Methicillinresistant Staphylococcus aureus (MRSA) isolated from Swiss and imported raw poultry meat collected at retail level. *Schweiz Arch Tierheilkd.* 158, 451-456.

Zurfluh, K., Wang, J., Klumpp, J., Nuesch-Inderbinen, M., Fanning, S. and Stephan, R. 2014. Vertical transmission of highly similar bla CTX-M-1-harboring IncI1 plasmids in Escherichia coli with different MLST types in the poultry production pyramid. *Front Microbiol.* 5, 519.

In: Beta-Lactamases: An Overview
Editor: Thomas E. Stead

ISBN: 978-1-53616-818-1
© 2020 Nova Science Publishers, Inc.

Chapter 2

RESISTANCE MECHANISMS IN *ACINETOBACTER* WITH AN EMPHASIS ON β-LACTAMASES: AN UPDATE

Aparna Shivaprasad[1,2], PhD and Beena Antony[1,3,], PhD*

[1]Department of Microbiology
[2]MV Jayaram Medical College, Bengaluru, India
[3]Fr. Muller Medical College, Mangaluru, India
Rajiv Gandhi University of Health Sciences, Karnataka, India

ABSTRACT

A. baumannii has emerged as a ubiquitous 'Superbug' causing both community and health care associated infections in the recent past. More than 40 genospecies of Acinetobacter have been identified using Multilocus sequence analysis. Among them *A. baumannii, A. pittii, A. nosocomialis, A. seifertii* and *A. dijkshoorniae* that forms the AB complex are more clinically relevant. A wide array of antimicrobial resistance mechanisms have been described in *A. baumannii*. The most concerning ones being the β-lactamases with carbapenemase activity, i.e., Ambler

---

[*] Corresponding Author's Email: beenafmmc@gmail.com.

class-D serine oxacillinases (OXA-type) and Ambler class-B metallo-β-lactamases (MBLs). More recently, *A.baumannii* is also known to possess New Delhi Metallo β-lactamase (NDM), a broad-spectrum MBL, renamed as Plasmid encoded carbapenemase resistant metallo β-lactamase (PCM).

In our personal experience of analysing 888 strains of *Acinetobacter* isolated from various clinical samples over a period of 5 years, 80.5%, were extremely drug resistant *A. baumannii* (XDR-AB), showing resistance to multiple classes of antibiotics including carbapenems and 75% of them produced biofilm. Molecular characterization of 251 isolates done by PCR revealed genes such as bla_{OXA-51} (90.4%), bla_{OXA-58} (58.2%), bla_{OXA-23} (57.4%). 41% of our isolates were NDM-1 producers. Co-existence of multiple MBL and carbapenemase genes in clinical isolates of *A. baumannii* is a potential threat in hospitals and is a problem to reckon with. It should be seriously considered and addressed with alternative and newer therapeutic strategies, strict infection control measures and continuous surveillance. Initial screening of the putative carbapenemase producers will help to organize intervention and early directed therapy.

Keywords: β-lactamases, AB complex, XDR-AB, MBL, Carbapenemases, NDM

INTRODUCTION

A. baumannii has emerged as a ubiquitous 'Superbug' causing both community and health care associated infections in the recent past. It is a predominant pathogen in many hospitals and is capable of causing infections anywhere in the body, thus adding extra furrows on the forehead of medical community worldwide. More than 40 genospecies of *Acinetobacter* have been identified using Multilocus sequence analysis. Four of the species, i.e., *A. calcoaceticus*, *A. baumannii*, A. *pittii* (formerly genomic species 3) and *A. nosocomialis* (formerly genomic species 13TU) are very closely related and difficult to distinguish from each other by phenotypic properties. Therefore, it was proposed to refer these species as *Acinetobacter calcoaceticus-Acinetobacter baumannii* (ACB) Complex (Peleg et al., 2008).

A. pittii and *A. nosocomialis* are clinically relevant as they have been implicated in the vast majority of both community-acquired and nosocomial infections but they cannot be differentiated by routine diagnostic tests from *A. baumannii*. They are often just referred to as *Acinetobacter baumannii* in the literature unless stated otherwise. *A. calcoaceticus* is an environmental species and never been implicated in clinical disease. Hence, the designation ACB complex may not be appropriate if used in clinical context. The term "*A. baumannii* complex (AB Complex)" with three clinically relevant species i.e., *A. baumannii, A. pittii* and *A. nosocomialis* might be considered more suitable, according to the recent classification. Recently, two new species, *Acinetobacter seifertii* (closely related to *A. nosocomialis*) and *Acinetobacter dijkshoorniae* (closely related to *A. pittii*) have been included with in AB complex (Nemec et al., 2015; Cosgaya et al., 2016). Currently AB complex consists of five *Acinetobacter* species associated with human diseases.

A. baumannii has the potential to respond swiftly to changes in selective environmental pressure. Upregulation of innate resistance mechanisms and acquisition of foreign determinants are the critical skills that have made *A. baumannii* the most troublesome pathogen. It has high levels of intrinsic resistance to antibiotics and in addition, has acquired resistance genes to virtually all antibiotics, including carbapenems. The enzymes such as Extended spectrum β-lactamases (ESBLs) conferring resistance to broad spectrum cephalosporins, carbapenemases to carbapenems, 16S rRNA methylase to all clinically relevant aminoglycosides and NDM/PCM showing resistance to all aminoglycosides, macrolides, sulfamethoxazole and carbapenems are also causes of concern. There are reports of resistance to fluoroquinolones, polymyxins (colistin) and tigecycline that has led to pan drug-resistance (Peleg et al., 2008). It resists desiccation and persists on inanimate abiotic surfaces including catheters, ventilators and other medical devices used in health care facilities for several months by forming biofilms that enhances bacterial transmission. Many of the isolates causing outbreaks are suspected phylogenetically to be closely related with I to III clonal groups. Strong biofilm formation is a part of pathogenic strategies of this bacterium (Pour et al., 2011). It has excellent colonizing potential and causes intermittent and endemic outbreaks. Once the focus is established, this opportunistic pathogen is difficult to eradicate. The elimination of the identified source often requires multiple interventions.

In a personal experience of analysing the prevalence of metallo β-lactamase (MBL) and Carbapenemase producing nosocomial multidrug resistant *Acinetobacter* species in Mangaluru, Karnataka State, South India, we isolated 888 strains of *Acinetobacter* over a period of 5 years. Of these, 72.3% of the isolates were *A. baumannii*, 22.3% were *A.nosocomialis* and 5.4% were *A. pittii*. 80.5% of the AB complex strains isolated from various clinical samples were all XDR-AB, showing resistance to multiple classes of antibiotics including carbapenems. 75% of them produced biofilm indicating that they were established nosocomial pathogens.

MECHANISM OF ANTIBIOTIC RESISTANCE

Acinetobacter is a known reservoir of multiple plasmids carrying antibiotic resistance markers and harbours multiple mechanisms of drug resistance. A wide array of antimicrobial resistance mechanisms have been described in *A. baumannii* which is either enzymatic or non-enzymatic. However, multiple mechanisms often work in concert to produce the same phenotype.

Non-Enzymatic Mechanisms

This includes alterations in the outer membrane protein (OMPs), changes in the affinity or expression of penicillin-binding proteins (PBPs) and increased activity of multidrug efflux systems/pumps such as AdeABC, RND-type, etc.

Enzymatic Mechanisms

The most prevalent mechanism of β-lactam resistance is enzymatic degradation by β-lactamases. *A. baumannii* possess all classes of β-lactamases (Bush, Jacob and Medeiros, 1995; Camp and Tatum, 2013). However, β-lactamases of the most concern are those with carbapenemase activity, i.e., Ambler class-D serine oxacillinases (OXA-type) and Ambler class-B metallo-β-lactamases (MBLs). Emergence of Carbapenemases and MBLs have been reported from many countries such as Europe, Asia, Australia and South America, as well as from different parts of Indian subcontinent, particularly in MDR nosocomial pathogens but not in USA (Baranak et al., 2002; Lee et al., 2003; Mendiratta et al., 2002; Mary et al., 2005; Navaneeth et al., 2002; Jayakumar et al., 2007; Aparna et al., 2017; Almasaudi, 2018).

A. baumannii possess an inherent class-D β-lactamase gene (bla$_{OXA-51}$ like) that has the ability to confer resistance to carbapenems. It naturally occurs in this bacterium, has chromosomal location and is globally present. Its product has more affinity for imipenem than for meropenem. Its role in carbapenem resistance is associated with the presence of ISAba1 (Pleg et al., 2008). Additionally, mechanisms of carbapenem resistance have emerged because of importation of distantly related class-D β-lactamase genes bla$_{OXA-23}$, bla$_{OXA-58}$ and bla$_{OXA-40/24}$. Analysis of the molecular epidemiology of carbapenem resistant *A. baumannii* in various parts of the world indicated a considerable degree of geographic specificity (Peleg et al., 2005; Poirel and Nordmann 2006).

Though less commonly identified in AB complex than OXA-type carbapenemases, the hydrolytic activity of metallo-β-lactamases (MBLs) is more potent (100-1000 fold). Unlike OXA-type genes, MBL genes are commonly found within class-1 integrons. Presence of MBLs has been observed in recent years throughout the world. Out of five MBLs, only three have been identified in AB complex i.e., IMP, VIM and SIM. Analysis of the molecular epidemiology of carbapenem resistant *A. baumannii* in various parts of the world indicated a considerable degree of geographic specificity. Clinical *A. baumannii* with MBLs has been identified in Europe,

Asia and South America but not in US. Many Indian studies have also been reported the detection of MBL producers. Several geographic regions have shown the presence of both OXA and MBL-type enzymes in the same *A. baumannii* strains (Peleg et al., 2008; Johnson and Woodford 2013; Martins et al., 2014; Cherkaoui et al., 2015; Tada et al., 2015).

Table 1. Genes conferring antibiotic resistance and resistance mechanisms in *Acinetobacter baumannii*

Enzyme Group, Gene Name	Description	Antibiotic resistance
β-lactamase genes		
ADC	Chromosomally integrated cephalosporinase	Extended spectrum cephalosporins
VIM	Acquired metallo-β-lactamase	All β-lactams except monobactams, evades all β-lactamase inhibitors
IMP	Stronger carbapenem-hydrolyzing activity than OXA	Carbapenem resistance
OXA	A group of carbapenem-hydrolyzing oxacillinases	Carbapenem resistance
TEM	A broad-spectrum enzyme	Narrow-spectrum cephalosporins, all penicillins except temollin
SHV	Plasmid-mediated. Includes SHV-1 & at least 23 variants	Extended-spectrum cephalosporins, ampicillin
Aminoglycoside-Modifying Enzymes (AME) genes		
aadB	Enzymatic inactivation by adenylation	Kanamycin, tobramycin, gentamicin
aacC1	Enzymatic inactivation by acetylation	Gentamicin resistance
aacC2	Enzymatic inactivation by acetylation	A number of aminoglycosides
aphA6	Enzymatic inactivation by phosphorylation	Kanamycin, neomycin, gentamicin, amikacin, etc
aadA1	Modifies the 3"-hydroxyl position of streptomycin & 9"-hydroxyl position of spectinomycin	Streptomycin & spectinomycin
Gene-encoding efflux pumps		
adeABC	Composed of AdeA, AdeB & AdeC proteins	Aminoglycosides, Quinolones, Tetracyclines & Trimethoprim
Point mutations		
gyrA	Point mutation at Ser83	Quinolones
parC	Point mutation at Ser80	Quinolones

* Courtesy: Camp C & Tatum OL, 2013.

AmpC β-lactamases are molecular class-C Cephalosporinases encoded on the chromosome. They are inherent to *A. baumannii* strains and also known as Acinetobacter derived cephalosporinases (ADCs). These enzymes are inducible by the presence of β-lactams (especially Cefoxitin, Imipenem and Clavulanic acid). These enzymes can hydrolyze primary and extended spectrum Penicillins and Cephalosporins including Cephamycins and Monobactams (Liu and Lin 2015). They are poorly inhibited by β-lactamase inhibitors such as Clavulanic acid. The bacterium stops producing the β-lactamase when the inducing agent is removed. Plasmid encoded AmpC enzymes are extremely rare and are almost always expressed constitutionally. AB complex also possess Extended Spectrum β-lactamases (ESBLs) of Ambler class-A group such as CTX-M, VEB, PER, TEM, SHV, etc (Naas et al., 2006; Bonnin et al., 2013; Sacha et al., 2012; Yezli et al., 2015).

A. baumannii is also known to possess New Delhi Metallo β-lactamase (NDM). It is a broad-spectrum β-lactamase (an MBL) which has been renamed now as Plasmid encoded carbapenemase resistant metallo β-lactamase (PCM). It can inactivate all β-lactams except aztreonam. NDM/PCM producers show resistance to all aminoglycosides, macrolides, sulfamethoxazole and carbapenems. Molecular studies coupled with conjugation experiments have confirmed that bla_{NDM-1} gene is located on transferable plasmids. It was reported for the first time in *Klebsiella pneumoniae* isolated from urine of a 59 year old male patient of Indian descent (Johnson and Woodford 2013). Since then, high incidence of NDM/PCM producers belonging to different genera of gram negative bacteria including *Acinetobacter* have been reported globally (Tada et al., 2015; Harbak et al., 2012; Bonnin et al., 2012; Bogaerts et al., 2012; Decousser et al., 2013; Revanthi et al., 2013; Aparna et al., 2017). Series of further variants of NDM-1, i.e., NDM-2 to 7 have also been identified (Espinal et al., 2011).

Along the course of time, *A. baumannii* has gained virulence determinants which have made it an all-rounder in the nosocomial field. The characteristics or factors that had made *A. baumannii* the most effective nosocomial pathogen includes AbaR resistance islands, biofilm formation, β-lactamase production, efflux pumps and outer membrane proteins that helps in adhesion. Whole genome sequencing of clinical epidemic *A. baumannii* strain has shown a 86 kb resistance island known as AbaR1, which is one of the largest described till now. Overall, 52 resistance genes have been identified out of which 45 (86.5%) were localized in AbaR1 resistance island (Liu et al., 2014). Prevention of dissemination of multidrug resistance in *A. baumannii* is not an easy task.

Many surveillance studies have been conducted throughout the world to demonstrate the prevalence of multiple drug resistance genes in clinical strains of *A. baumannii* and the need of infection control measures including antibacterial management and prompt identification of β-lactamase producing isolates (Peleg et al., 2008; Cherkaoui et al., 2015; Hutheesing 2011; Mataseje et al., 2012; Noori et al., 2014). Genes conferring antibiotic resistance & resistance mechanisms in *Acinetobacter baumannii* is shown in Table 1. It is critical to document and understand the ability, dissemination pathways and treatment patterns of *A. baumannii* clones for proper patient management.

PHENOTYPIC METHODS TO DETECT BETA-LACTAMASES

The multi-drug resistant and Pan resistant *A. baumannii* is threatening the current antibiotic era. Detection of MBL producing multi-drug resistant *A.baumannii* (MDR AB) is crucial for the optimal and modified therapy and also to initiate effective control of dissemination of resistance. Various methods such as Imipenem-EDTA Double-Disc Synergy Test (DDST) and Imipenem-EDTA Combined Disc test (CDT) (Yong et al., 2002), Modified Hodge's test (MHT) (Lee et al., 2003), Microdilution test, MBL E-Test (Walsh et al., 2002) are available for MBL and carbapenemase detection. In an attempt to detect β-lactamase producing strains of *A.baumannii* among clinical isolates by various phenotypic tests, the results we obtained and their comparative evaluation is shown in Figures 1 to 4.

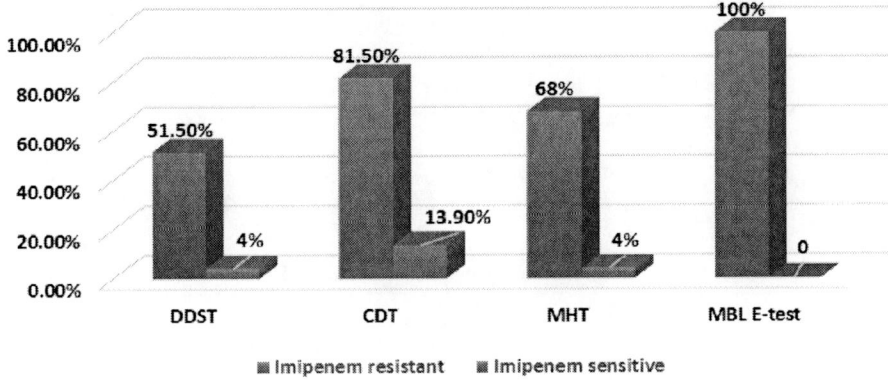

Figure 1. Comparison of efficiency of Phenotypic tests for MBL & Carbapenemase detection.

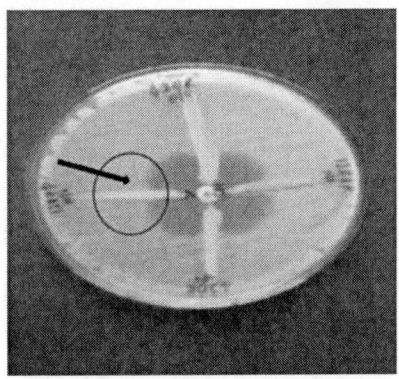

Figure 2. Positive Modified Hodge test showing distorted zone (Clover-leaf shaped zone of inhibition).

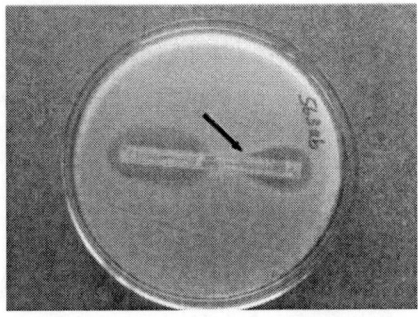

Figure 3. Positive MBL E-test.

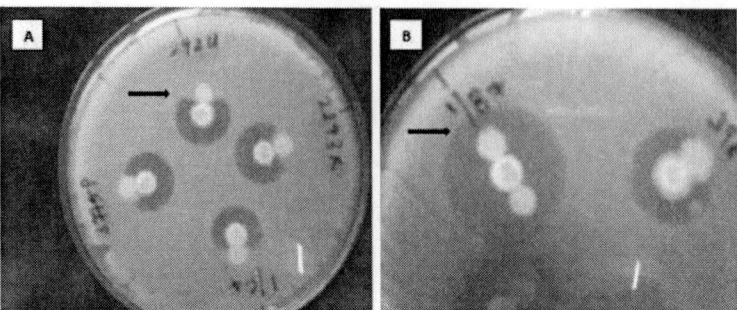

Figure 4. A - Positive AmpC disc test showing indentation of the zone of inhibition; B - Negative test.

Simple and accurate tests are needed to detect MBL and carbapenemase producers in routine diagnostic laboratory. As there are no standard guidelines available for the detection of these enzymes, considerable disagreement exist in the literature regarding the best technique for their phenotypic detection. Several studies have reported the use of DDST as one of the convenient methods for the detection of Ambler class-B MBL production. However, in our study DDST showed the lowest positivity (51.5%) and our result was similar to the study conducted by few other workers (John and Balagurunathan 2011; Behra et al., 2008). The disadvantage of DDST was subjective interpretation of result in some instances. Alternatively combined disc test was found to be more efficient as EDTA is added to imipenem disc instead of blank disc as done in DDST. Results of our study have shown that the positivity rate is higher in CDT which ranged from 60.9% to 100%, suggesting that it is an efficient method to detect MBL in routine diagnostic laboratory and far superior to DDST as well, which is in accordance with the findings of other investigations (Martins et al., 2014; Bonnin et al., 2013; Behra et al., 2008; Amudhan et al., 2011; Chaudhary and Payasi 2012; Rynga et al., 2015; Aparna et al., 2014).

CLSI does not advocate the use of Modified Hodge test for detection of carbapenemase production in Non-fermenting Gram negative bacilli and it alone may not confirm NDM producing microorganisms. Still several authors have found the use of MHT with Imipenem as a useful screening test for carbapenemase production and for NDM screening but it may require complementary tests such as use of inhibitors. Synergistic effect by using chemical and biological inhibitor along with an indicator organism helps in confirming NDM production and allows systematic categorization of *A. baumannii* strains according to their carbapenem susceptibility and genotyping tests (John and Balagurunathan 2011; Azimi et al., 2015, Wong et al., 2015). Even though some of the previously published data have shown low positivity of MHT, many studies have reported high positivity rate ranged from 96.7% to 100% (Amudhan et al., 2011; Sofiane et al., 2015; Niranjan et al., 2013; Karthikeyan et al., 2010; Hasan et al., 2014), whereas our study showed moderate positivity and sensitivity (68%).

E-test, a quantitative technique for determining the MIC of antimicrobial agents against microorganisms, is a very sensitive (100%) method for detection of MBLs in MDR bacteria including *Acinetobacter* as implied in published literature including our data (Niranjan et al., 2013; Sofiane et al., 2014; Yong et al., 2015; Aparna et al., 2014), however expensive.

Various MBL inhibitors such as Ethylene diamine tetra acetic acid (EDTA), 2-mercaptoacetic acid (MAA) and 2-mercaptopropionic acid (MPA) have been employed by different investigators. Some studies report that MPA provides higher sensitivity, although interpretation is difficult and subjective (Lee et al., 2003; Martins et al., 2014), while few other studies suggest that EDTA could be used successfully. Based on the results obtained in majority of the published reports, EDTA is considered as the best MBL inhibitor for routine use (Mendiratta et al., 2002; Mary et al., 2005; Navaneeth et al., 2002; Jayakumar et al., 2007; Peleg et al., 2005; Aparna et al., 2014). The action of EDTA on the permeability of the cell wall can accelerate the breakdown of imipenem and can enhance the expression of membrane proteins (Martins et al., 2014).

Simultaneous existence of different beta-lactamases in clinical isolates is a problem to reckon with and should be seriously considered for alternative and newer therapeutic strategies, strict infection control measures and continuous surveillance. Easy detection methods are required for each of these in routine clinical laboratories. Phenotypic detection of MBLs and carbapenemases in routine diagnostic laboratories can be effectively done by using CDT, MHT and MBL E-test. ESBL and AmpC enzymes can be detected by Phenotypic Confirmatory Disc Diffusion Test (PCDDT) and AmpC Disk test. They are easy to perform, economical and helps in monitoring the emergence MBLs, carbapenemases and other β-lactamases in MDR AB including NDM/PCM screening. Initial screening of the putative carbapenemase producers will help to organise intervention and early directed therapy.

GENOTYPIC METHOD TO DETECT BETA-LACTAMASES

Nosocomially acquired MDR AB pose a real challenge to the clinician. Since 2000, carbapenem-resistant and multidrug-resistant *A. baumannii* strains (MDR and XDR strains) have emerged due to the presence of multiple carbapenemases (Peleg et al., 2008; Camp and Tatum 2013: Niranjan et al., 2013). One of the most striking feature of *A. baumannii* is its extraordinary ability to develop resistance against major classes of antibiotics and have been found to harbour multiple resistance genes against carbapenems (Peleg et al., 2008; Niranjan et al., 2013). Molecular tests and comparative genomics analysis offer insight to the mechanisms of resistance and virulence in *A. baumanni*. It is necessary to understand the repertoire of resistance determinants and their organization and origins to explain their acquisition and dissemination. Molecular characterization of suspected pathogens might help to provide appropriate antimicrobial therapy and good clinical outcome. Hence, our study also included detection and prevalence of resistance genes in *Acinetobacter* isolates by Polymerase Chain Reaction (Figure 5). All our isolates were XDR-AB, however, Pan drug resistant isolates were not seen in our hospital. Maximum of them (90.4%) were *A. baumannii* as they had bla_{OXA-51} gene (intrinsic carbapenemase gene present in chromosome) and the remaining 9.6% comprised of *A. pittii* and *A. nosocomialis*.

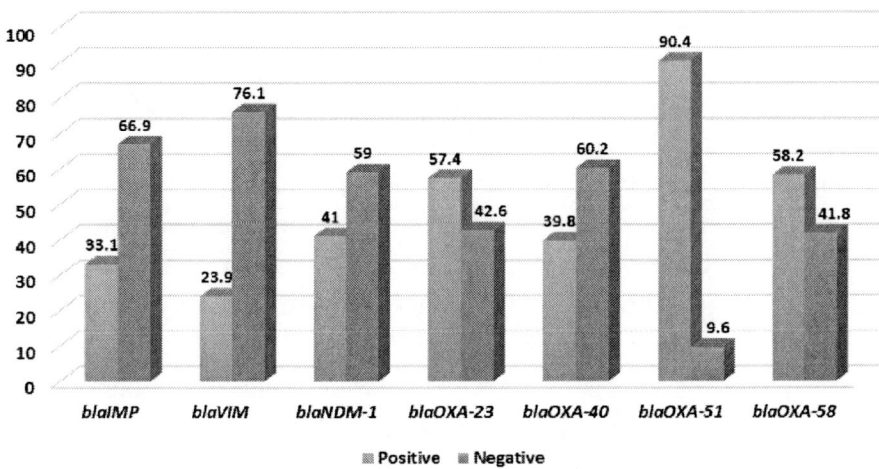

Figure 5. Distribution of MBL & Carbapenemase genes *Acinetobacter* isolates.

Table 2. Differences in the antimicrobial resistance mechanisms among the members of AB complex

Mechanism	A. baumannii	A. pittii	A. nosocomialis
Resistance to carbapenems –			
• Class B – MBLs	++	++++	++++
• Class D – Oxacillinases	++++	+	-
• bla_{OXA-51} like gene	Intrinsic	-	-
Genes encoding Aminoglycoside modifying enzyme	$armA$ & $aph(3´)$-Ia	-	$aph(3´)$-VI
Resistance to fluoroquinolones	Resistant Ser83Leu substitution in $gyrA$ gene	Susceptible Wild-type Ser83 in $gyrA$ gene	Susceptible Wild-type Ser83 in $gyrA$ gene
Efflux transporters –			
• AdeABC & AdeIJK	++++	-	-
• AdeDE & AdeXYZ	-	++++	-

The drug resistant international clones clone II (IC2) or sequence type 2 (ST2) were found to spread from other countries to Indian subcontinent (Rynga et al., 2015; Mahdian et al., 2015). Few studies have reported that outbreaks were usually caused by ST1, ST2, ST15, however, Ou, et al., reported a ST417 strain and an unusual, hypervirulent, extremely rare MDR MLST type ST10 *A. baumannii* (LAC-4) strain that caused clinical outbreak (Ou, et al., 2015). Worldwide literature survey highlights the co-existence of various resistance genes in MDR and XDR *A. baumannii* isolates. The co-production of OXA and metallo-β-lactamase enzymes is not an uncommon phenomenon in *A. baumannii* (Peleg et al., 2008; Chen et al., 2011; Johnson and Woodford 2013; Martins et al., 2014; Tada et al., 2016; Niranjan et al., 2013; Chaudhary and Payasi 2012; Karthikeyan et al., 2010). The results of molecular analysis of our study was in agreement with reports of all those published studies about co-existence of multiple resistance genes except for geographical variation (Aparna et al., 2017). *bla*$_{OXA-51}$ is involved in intrinsic resistance with chromosomal origin whereas the other genes including *bla*$_{NDM-1}$ causes resistance via plasmids which is responsible for the increased resistance spectrum by horizontal transfer of resistance factors in intra- and inter-species (Poirel and Nordmann 2006; Hasan et al., 2014). Investigations of the present study revealed that, 30% clinical isolates of XDR *Acinetobacter baumannii*, carried four genes and three isolates (1.2%) had all the seven genes which indicated the increased prevalence of resistance genes in XDR AB. Prevalence of these genes were found to be higher in our report compared to other studies as it was the first comprehensive study which probed for higher number of multiple genes, i.e., seven resistance genes – three MBLs including the novel NDM/PCM and four oxacillinases in clinical isolates of *A. baumannii*. Our findings also happened to be the first to report the highest prevalence of the novel *bla*$_{NDM-1}$ (41%) gene from south India. It was not possible to find any correlation between the genotypic profile and susceptibility pattern but the difference might be because of other non-enzymatic mechanisms that exists in *A. baumannii* and also the co-presence of ESBL or AmpC β-lactamases that might have rendered carbapenemase ineffective against such isolates (Aparna et al., 2014; Aparna et al., 2017). Our experience also suggested that the production of MBL and OXA-type of carbapenemases were the predominant mechanism(s) of carbapenem resistance in strains isolated from southern India which is in accordance with the previously published studies. As reported by various investigators (Saranath et al., 2015; Cherkaoui et al., 2015; Tada et al., 2015), the antimicrobial resistance mechanisms are distinct for the AB complex (Table 2).

SEQUENCING OF *A. BAUMANNII* RESISTANCE GENES

A. baumannii undergoes rapid genome evolution within a hospital outbreak. There is considerable genotypic diversity within patient associated and environmental isolates of *Acinetobacter*. By sequencing the genome, it is possible to link patient-derived isolates directly to environmental isolates which in turn helps to identify the environmental source and its eradication (Halachev et al., 2014). The comparative genomic analysis done by Liu, et al., 2014 revealed the existence of extensive genomic variation in *A. baumannii* genome. Results of their study proved that transposons, genomic islands and point mutations were the main contributors to the plasticity of *A. baumannii* genome and played a critical role in facilitating the development of antibiotic resistance in clinical isolates (Liu et al., 2014).

Whole genome sequencing (WGS) is now poised to make an impact on hospital infection prevention and control, delivering cost-effective identification of routes of infection within a clinically relevant time frame and allowing infection control teams to track and even prevent the spread of drug resistant nosocomial pathogens (Halachev et al., 2014). It provides a promising novel tool for investigating the epidemiology of outbreaks, particularly when coupled to clinical location and temporal data. In our setup, an attempt was done to investigate the sequencing of more prevalent resistance genes such as bla_{OXA-51}, bla_{OXA-58}, bla_{OXA-23} and bla_{NDM-1} to check for any mutations or variations. Bioinformatics of sequencing of the bla_{OXA-51}, bla_{OXA-58} and bla_{NDM-1} showed variations in the gene sequence when compared to the original Gen bank sequence indicating that our isolates were different and had undergone mutation. There was not much difference in bla_{OXA-23} gene sequence in our isolates. An interesting observation was made in this study that 20 of the 49 imipenem sensitive strains, showed the presence of two or more resistance genes, however, negative by all the four phenotypic tests. This finding suggests that in spite of the presence of the genes, absence of the phenotypic expression could be a cause of concern as these isolates might help in dissemination of these resistance genes. It also implies that the scenario might change in due course and these isolates might start producing enzymes, becoming resistant to carbapenems. Hence both phenotypic screening and genotyping should be performed for the detection of resistance genes in nosocomial MDR and XDR *A. baumannii* isolates (Aparna et al., 2017).

Conclusion

MDR *A. baumannii* complex is the most troublesome pathogen in nosocomial infections. The co-existence of multiple carbapenemases and MBLs along with other resistance mechanisms might result in treatment failure. Molecular characterization of suspected pathogens might help to provide appropriate antimicrobial therapy and good clinical outcome. *A. baumannii* isolates harbouring multiple resistance genes should be seriously considered and addressed with newer therapeutic strategies, strict infection control measures and continuous surveillance. Initial screening of the putative carbapenemase producers would help to organize intervention and early directed therapy. The detection and analysis of predominant genotype among the MDR strains in hospitals not only reinforces the importance of intervention measures such as contact precautions but also helps to reduce the chances of cross infection, to better match the physical areas of hospitals/clinics and to improve nosocomial infection control.

References

Almasaudi, S. B. (2018). Acinetobacter species as nosocomial pathogens: epidemiology and resistance features. *Saudi Journal of Biological Sciences*, 25:586-596.

Amudhan, S. M., Sekar, U., Arunagiri, K. and Sekar, B. (2011). OXA beta-lactamase-mediated carbapenem resistance in *Acinetobacter baumannii*. *Indian Journal of Medical Microbiology*, 29:269-274.

Aparna, S., Beena, A. and Poornima, K. S. (2014). Comparative evaluation of four phenotypic tests for detection of metallo-β-lactamase and carbapenemase production in *Acinetobacter baumannii*. *Journal of Clinical and Diagnostic Research*, 8:DC 05-08.

Aparna, S., Beena, A. and Chaitra C. (2017). Genotyping of MDR *Acinetobacter baumannii* to detect the co-existence of MBL and carbapenemase genes including NDM-1 and the sequencing of the most prevalent resistance genes. (2017). *Indian Journal of Applied Research*, 7 (10):10-13.

Azimi, L., Lari, A. R., Talebi, M., et al., (2015). Inhibitory-based method for detection of *Klebsiella pneumoniae* carbapenemase *Acinetobcter baumannii* isolated from burn patients. *Indian Journal of Pathology and Microbiology*, 58:192-194.

Baranak, A., Fiett, J., Sulikowsha, A., et al., (2002). Country wide spread of extended spectrum β-lactamases producing microorganisms of the family *Enterobacteriaceae* in Poland. *Antimicrobial Agents and Chemotherapy*, 46:151-159.

Behra, B., Mathur, P., Das, A., et al., (2008). An evaluation of four different phenotypic techniques for detection of Metallo-β-Lactamase producing *Pseudomonas aeruginosa*. *Indian Journal of Medical Microbiology*, 26:233-237.

Bogaerts, P., de Castro, R. R., Roisin, S., et al., (2012). Emergence of NDM-1-Producing *A. baumannii* in Belgium. *Journal of Antimicrobial Chemotherapy*, 66:1552-1553.

Bonnin, R. A., Poirel, L., Naas, T., et al., (2012). Dissemination of New Delhi Metallo-β-Lactamase-1 producing *Acinetobacter baumannii* in Europe. *Clinical Microbiology and Infection*, 18:E362.

Bonnin, R. A., Rotimi, V. O., Al Hubail, M., et al., (2013). Wide dissemination of GES-Type Carbapenemases in *Acinetobacter baumannii* isolates in Kuwait. *Antimicrobial Agents and Chemotherapy*, 57:183-188.

Bush, K., Jacoby, G. A. and Medeiros, A. A. (1995). A Functional classification scheme for β-lactamases and its correlation with molecular structure. *Antimicrobial Agents and Chemotherapy*, 39: 1211-1233.

Camp, C. and Tatum, O. L. (2013). A Review of *Acinetobacter baumannii* as a highly successful pathogen in times of war. *Laboratory Medicine*, 41:650-657.

Chaudhary, M. and Payasi, A. (2012). Molecular characterization and antimicrobial susceptibility study of *Acinetobacter baumannii* clinical isolates from Middle East, African and Indian patients. *Journal of Proteomics and Bioinformatics*, 5:265-269.

Chen, Z., Qiu, S., Wang, Y., et al., (2011). Coexistence of bla$_{NDM-1}$ with the prevalent bla$_{OXA-23}$ and bla$_{IMP}$ in Pan Drug Resistant *Acinetobacter baumannii* isolates in China. *Clinical Infectious Disease*, 52:692-693.

Cherkaoui, A., Emonet, S., Renzi, G. and Schrenzel, J. (2015). Characteristics of multidrug-resistant *Acinetobacter baumannii* strains isolated in Geneva during colonization or infection. *Annals of Clinical Microbiology and Antimicrobials*, 14:42.

Cosgaya, C., Mari-Almirall, M., Van Assche A., et al., (2016). Acinetobacter dijkshoorniae sp. nov., a member of *Acinetobacter calcoaceticus-Acinetobacter baumannii* complex mainly recovered from clinical samples in different countries. *International Journal of Systemic and Evolutionary Microbiology*, 66(10): 4105-4111.

Decousser, J. W., Jansen, C., Nordmann, P., et al., (2013). Outbreak of NDM-1-producing *Acinetobacter baumannii* in France, January to May 2013. *European Surveillance*, 18:pii=20547.

Espinal, P., Fugazza, G., Lopez, Y., et al., (2011). Dissemination of an NDM-2-producing *Acinetobacter baumannii* clone in an Israeli rehabilitation centre. *Antimicrobial Agents and Chemotherapy*, 55:5396-5398.

Halachev, M. R., Chan, J. Z., Constantinidou, C. I., et al., (2014). Genomic epidemiology of a protracted hospital outbreak caused by multidrug-resistant *Acinetobacter baumannii* in Birmingham, England. *Genome Medicine*, 6:70.

Harbak, J., Stolbova, M., Studentova, V. et al., (2012). NDM-1 producing *Acinetobacter baumannii* isolated from a patient repatriated to the Czech Republic from Egypt, July 2011. *European Surveillance*, 17:pii=20085.

Hasan, B., Perveen, K., Olsen, B. and Zahra, R. (2014). Emergence of Carbapenem-Resistant *Acinetobacter baumannii* in Hospitals in Pakistan. *Journal of Medical Microbiology*, 63:50-55.

Hutheesing, N. (2011). Eight deadly bugs lurking in hospitals. *African Traditional Herbal Research Clinic*, 6:8-1.

Jayakumar, S. and Appalaraju, B. (2007). Prevalence of multi and pan drug resistant *Pseudomonas aeruginosa* with respect to Extended Spectrum β-lactamases and Metallo β-lactamase in a tertiary care hospital. *Indian Journal of Pathology and Microbiology*, 50:922-925.

John, S. and Balagurunathan, R. (2011). Metallo beta lactamase producing *Pseudomonas aeruginosa* and *Acinetobacter baumannii*. *Indian Journal of Medical Microbiology*, 29:302-304.

Johnson, P. A. and Woodford, N. (2013). Global spread of antibiotic resistance: the example of New Delhi Metallo-β-Lactamase (NDM)-mediated carbapenem resistance. *Journal of Medical Microbiology*, 62:499-513.

Karthikeyan, K., Thirunarayanan, M. A. and Krishnan, P. (2010). Coexistence of bla_{OXA-23} with bla_{NDM-1} and armA in clinical isolates of *Acinetobacter baumannii* from India. *Journal of Antimicrobial Chemotherapy*, 65:2253-2254.

Lee, K., Lim, Y. S., Ynog, D., et al., (2003). Evaluation of the Hodge Test and the Imipenem-EDTA Double Disk Synergy Test for differentiation of metallo β-lactamases producing clinical isolates of *Pseudomonas* spp and *Acinetobacter* spp. *Journal of Clinical Microbiology*, 41:4623-4629.

Liu, F., Zhu, Y., Yi, Y., et al., (2014). Comparative genomic analysis of *Acinetobacter baumannii* clinical isolates reveals extensive genomic variation and diverse antibiotic resistance determinants. *BMC Genomics*, 15:3058-3085.

Liu, Y. and Lin, X. (2015). Detection of AmpC (beta)-lactamases in *Acinetobacter baumannii* in the Xuzhou region and analysis of drug resistance. *Experimental Therapeutic Medicine*, 10:933-936.

Mahdian, S., Sadeghifard, N., Pakzad, I., et al., (2015). *Acinetobacter baumannii* clonal lineages I and II harbouring different carbapenem-hydrolyzing-β-lactamase genes are widespread among hospitalized burn patients in Tehran. *Journal of Infections and Public Health*, 8:533-542.

Martins, H. I. S., Bomfim, M. R. Q., Franca, R. O., et al., (2014). Resistance markers and genetic diversity in *Acinetobacter baumannii* strains recovered from nosocomial blood stream infections. *International Journal of Environmental Research and Public Health*, 11:1465-78.

Mary, V. J., Kandathil, A. J. and Balaji, V. (2005). Comparison of methods to detect carbapenemase and metallo β-lactamase production in clinical isolates. *Indian Journal of Medical Research*, 121:780-783.

Mataseje, L. F., Bryce, E., Roscoe, D., et al., (2012). Carbapenem-resistant gram negative bacilli in Canada 2009-10: Results from the Canadian Nosocomial Infection Surveillance Program (CNISP). *Journal of Antimicrobial Chemotherapy*, 67:1359-1367.

Mendiratta, D. K., Deotale, V. and Narang P. (2002). Metallo β-lactamase producing *Pseudomonas aeruginosa* in a hospital from a rural area. *Indian Journal of Medical Research*, 121:701-703.

Naas, T., Coignard, B., Carbonne, A., et al., (2006). VEB-1 Extended-Spectrum β-lactamase producing *Acinetobacter bauamnnii*, France. *Emerging Infectious Diseases*, 12:1214-1222.

Navaneeth, B. V., Sridaran, D., Sahay, D. and Belwadi M. R. S. (2002). A preliminary study on metallo β-lactamase producing *Pseudomonas aeruginosa* in hospitalized patients. *Indian Journal of Medical Research*, 116:264-7.

Nemec, A., Krizoa, L., Maixnerova., et al., (2015). *Acinetbacter seifertii* sp. nov., a member of *Acinetobacter calcoaceticus-Acinetobacter baumannii* complex isolated from human clinical specimen. *International Journal of Systemic and Evolutionary Microbiology*, 65 (3):934-942.

Niranjan, D. K., Singh, N. P., Manchanda, V., et al., (2013). Multiple carbapenem hydrolyzing genes in clinical isolates of *Acinetobacter baumannii*. *Indian Journal of Medical Microbiology*, 31:237-241.

Noori, M., Karimi, A., Fullah, F., et al., (2014). High prevalence of Metallo-β-Lactamase producing *Acinetobacter baumannii* isolated from two hospitals of Tehran, Iran. *Archives of Paediatric Infectious Diseases*, 2:e15439.

Ou, H. Y., Kuang, S. N., He, X., et al., (2015). Complete genome sequence of hypervirulent and outbreak-associated *Acinetobacter baumannii* strain LAC-4: Epidemiology, resistance genetic determinants and potential virulence factors. *Scientific Reports*, 5:8643.

Peleg, A. Y., Franklin, C., Bell, J. M. and Spelmann, D. W. (2005). Dissemination of the metallo-β-lactamase gene blaIMP4 among gram negative pathogens in a clinical setting in Australia. *Clinical Infectious diseases*, 41:1549-1556.

Peleg, A. Y., Seifert, H. and Paterson, D. L. (2008). *Acinetobacter baumannii*: Emergence of a successful pathogen. *Clinical Microbiological Review*, 21:538-582.

Poirel, L. and Nordmann, P. (2006). Carbapenem resistance in *Acinetobacter baumannii*: mechanisms and epidemiology. *Clinical Microbiology and Infection*, 12:826-836.

Pour, N. K., Dusane, D. H., Dhakephalkar, P. K., et al., (2011). Biofilm formation by *Acinetobacter baumannii* strains isolated from urinary tract infection and urinary catheters. *FEMS Immunology & Medical Microbiology*, 62:328-38.

Revanthi, G., Siu, L. K., Lu, P. L. and Huang, L. Y. (2013). First report of NDM-1 producing *Acinetobacter baumannii* in East Africa. *International Journal of Infectious Diseases*, 17:e1255-1258.

Rynga, D., Shariff, M. and Deb, M. (2015). Phenotypic and molecular characterization of clinical isolates of *Acinetobacter baumannii* from Delhi, India. *Annals of Clinical Microbiology and Antimicrobials*, 14:40.

Sacha, P., Wieczorek, P., Ojdana, D., et al., (2012). Susceptibility, phenotypes of resistance, and extended-spectrum (Beta)-lactamases in *Acinetobacter baumannii* strains. *Folia Histochemia et Cytobiologica*, 50:46-51.

Saranath, R., Vasanth, V., Vasanth, T., et al., (2015). *Acinetobacter baumannii* strains of CC103B and CC92B clonal complexes harbouring

OXA-type carbapenemases and metallo-β-lactamases in Sothern India. *Microbiology and Immunology*, doi:10.1111/1348-0421.12252.

Sofiane, B., Ahmed, A. S., Abdelaziz, T. and Jean-Marc, R. (2014). Characterization of *Acinetobacter baumannii* clinical isolates carrying bla_{OXA-23} carbapenemases and 16S rRNA methylase *arm*A genes in Yemen. *Microbial Drug Resistance*, 20:604-609.

Sofiane, B., Olumuyiwa, O. A., Houria, A., et al., (2015). Emergence of colistin- and carbapenem-resistant *A. baumannii* ST2 clinical isolates in Algeria: First Case Report. *Microbial Drug Resistance*, doi:10.1089/mdr.2014.0214.

Tada, T., Miyoshi-Akiyama, T., Shimada, K., et al., (2014). Dissemination of 16S rRNA methylase armA- producing *Acinetobacter baumannii* and emergence of oxa-72 carbapenemase coproducers in Japan. *Antimicrobial Agents and Chemotherapy*, 59:2916-2920.

Tada, T., Miyoshi-Akiyama, T., Shimada, K., et al., (2015). Dissemination of clonal complex 2 *Acinetobacter baumannii* strains co-producing carbapenemases and 16S rRNA methylase ArmA in Vietnam. *BMC Infectious Diseases*, 15:438.

Walsh, T. R., Bolmstrom, A., Qwarnstrom, A. and Gales, A. (2002). Evaluation of a new E-test for detecting metallo β-lactamase in routine clinical testing. *Journal of Clinical Microbiology*, 23:120-124.

Wong, M. H., Li, Y., Chan, E. W. and Chen, S. (2015). Functional categorization of carbapenemase-mediated resistance by a combined genotyping and two-tiered Modified Hodge test approach. *Frontier Microbiology*, 6:293.

Yezli, S., Shibl, A. M. and Memish, Z. A. (2015). The molecular basis of β-lactamase production in gram negative bacteria from Saudi Arabia. *Journal of Medical Microbiology*, 64:127-136.

Yong, D., Lee, K., Yum, J. H., et al., (2002). Imipenem-EDTA disc method for differentiation of metallo β-lactamases producing clinical isolates of *Pseudomonas* spp and *Acinetobacter* spp. *Journal of Clinical Microbiology*, 40:3798-3801.

Yong, Z., Xinwei, W., Xinqiang, Z., et al., (2015). Genetic characterization of ST195 and ST365 carbapenem-resistant *Acinetobacter baumannii* harbouring *bla*$_{OXA-23}$ in Guangzhou, China. *Microbial Drug Resistance*, doi:10.1089/mdr.2014.0183.

In: Beta-Lactamases: An Overview
Editor: Thomas E. Stead

ISBN: 978-1-53616-818-1
© 2020 Nova Science Publishers, Inc.

Chapter 3

GOLDEN AGE OF β-*LACTAM* ANTIBIOTICS TO ANTIBIOTIC RESISTANCE

*Ritika Chauhan and Jayanthi Abraham**
Microbial Biotechnology Laboratory, SBST,
VIT University, Vellore, Tamil Nadu

ABSTRACT

Beta-lactam antibiotics are a group of antibiotics that hold β-lactam ring in molecular structure. These are widely used antibiotics mainly in developing countries against bacterial infections. β-*lactam* group of antibiotics, originally found in fungi *Penicillium notatum* inhibit cell wall biosynthesis in bacteria especially Gram-positive organisms. The development of new derivatives of β-lactam class- carbapenem, cephalosporins, cephamycin, monobactam are efficient against bacterial species which have developed resistance mechanisms. This review will walk readers through the discovery and golden age of β-lactam, mechanism of action as broad-spectrum antibiotic, expansion of beta-lactam derivatives, β-lactamase inhibitors to antibiotic resistance in β-lactams.

---

* Corresponding Author's Email: jayanthi.abraham@gmail.com.

Keywords: beta-lactams, cephalosporin, carbapenem, antimicrobial resistance

INTRODUCTION

The discovery of Penicillin by Alexander Fleming in 1929, is the most significant and remarkable development in medicine. While working with *Staphylococcus* colonies, Alexander Fleming observed the lysis of *Staphylococcus* colonies on culture plate which was contaminated by white colour fluffy mold. He was able to grow the airborne mold at room temperature and identified it as *Penicillium rubrum* (Fleming, 1929). Fleming and his assistants Craddock and Ridley, extracted the lytic compound from the broth with acetone or alcohol to perform antimicrobial assay. The penicillin extracts inhibited the growth of *Staphylococci, Streptococcus pyogenes* and *Pseudomonas* whereas extracts were found inactive against *Escherichia coli, Haemophilus influenzae, Salmonella typhi, Pseudomonas aeruginosa, Bacillus proteus* or *Vibrio cholerae* (Hare, 1982). The potential of common airborne mold *Penicillium* as therapeutic agent was proved by Fleming, Florey and Abraham, when mold was smuggled to America to treat Allied soldiers in World War II and saved many lives. However, Fleming was not successful to purify penicillin from broth to conduct clinical study, the use of penicillin as therapeutic agent to treat bacterial infections did not happen until 1940. In August 1940, Florey, Chain and their co-workers reported the use of penicillin against infections in mice, rats and cats (Kong et al., 2010). These magic bullets set a stage for the development of industrial microbiology and most successful class of antibiotics-beta lactams (Macfarlane 1979; Bush and Bradford 2016).

β-lactams

β-lactam class of antibiotics are widely used antibiotics against bacterial infections over 70 years (Watkins and Bonomo 2017). The annual sale of β-lactam class of drugs is almost US $15 billion and accounts 65% of total international antibiotics market (Thakuria and Lahon 2013). The four membered β-lactam ring is an integral part of widely used antibiotics, has been currently classified into: Penicillins, Cephalosporins, Carbapenems, Monobactams, Cephamycins and β-lactamase inhibitor combinations. These antibiotics are efficacious and widely prescribed against Gram positive and Gram negative bacterial infections. The detailed classification of beta-lactam antibiotics is presented in Figure 1.

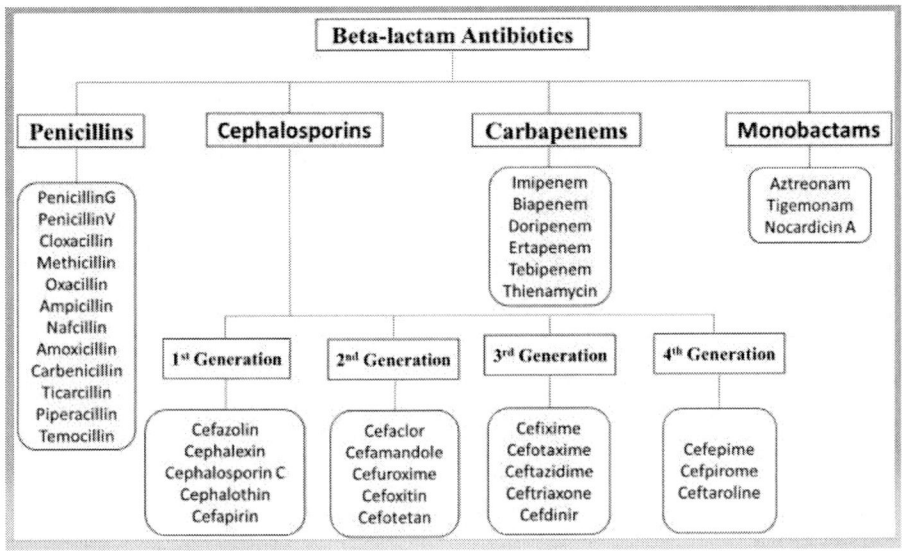

Figure 1. Classification and different generation of beta-lactam antibiotics.

The β-lactam antibiotics act by inhibiting bacterial cell wall biosynthesis-peptidoglycan layer and has lethal effect on bacterial population. Peptidoglycan layer is an important component of cell wall structure, synthesis of peptidoglycan is promoted by D-D-transpeptidases also known as Penicillin Binding Proteins (PBP)-group of enzymes characterized by their affinity towards Penicillin. The bacterial cell wall is comprised of *N-acetyl-muramic acid* (NAM) and *N-acetylglucosamine* (NAG) subunits, PBP binds to peptide chain forming a crosslink between adjacent glycan chains with release of D-alanine-D-alanine residue (Goffin and Ghuysen 1998; Sauvage et al., 2008). The covalent bonding between peptide and sugar chain, protects the bacterial cell from osmotic rupture-providing rigidity and stability to cell wall. When beta lactam class of antibiotics- Penicillins, Cephalosporins, Monobactams and carbapenems enters into host, β-lactam ring bears structural similarity to D-alanine-D-alanine residues, PBP mistakenly use β-lactam as building block for cell wall synthesis and tightly binds to transpeptidase active site. This results in slow synthesis of bacterial cell wall, peptidoglycan autolysis, increased permeability and breakdown of cell wall (Bayles 2000).

The β-lactam antibiotics have been classified into subgroups ranging from narrow spectrum, broad spectrum to extended spectrum antibiotics. These antibiotics are empirically used against various bacterial infections such as middle ear infections, respiratory & urinary tract infections, hospital associated *Pseudomonas* infection, meningitis, septicaemia and endocarditis (Thakuria and Lahon 2013). Among beta lactam antibiotics, cephalosporins are highly utilized and prescribed drug of choice against bacterial infections.

Penicillins

The benzylpenicillin (Penicillin G) was the first antibiotic clinically used to treat *Streptococcal* sepsis in 33-year-old woman, New Haven, Connecticut hospital, Yale, the patient was fully recovered from *Streptococcal* infection from her bloodstream with the help of "wonder drug" (Arias 2009). The benzylpenicillins are efficient inhibitors of Gram positive bacteria- *Staphylococci, Streptococci, Pneumococci, Bacillus anthracis, Clostridium perferinges, Corynebacterium diphtheriae* when compared to Gram negative bacteria (Mariya and Pilla 2017). The outer membrane of Gram negative bacteria acts as selective barrier for blocking the entry of penicillin. Another factor for inability of benzylpenicillins to act against Gram negative bacteria is penicillinases- beta lactamases, group of enzymes which hydrolysis β-lactam ring (Sutherland 1964). The phenoxymethylpenicillin (Penicillin V), naturally occurring penicillin, is an oral formulation used to treat mild to moderate *Streptococcal* infections especially in paediatric patients (Pottegard et al., 2014). The list of penicillin derivatives is presented in Table 1.

The unavoidable and continuous use of penicillin, resulted in penicillin resistant- penicillinase producing *Staphylococci* in patients. In 1942, four strains of *Staphylococcus aureus* were found to be resistant against Penicillin G in patients, within few years the infections caused by penicillin-resistant *Staphlococcus aureus* started spreading from hospitals to communities. This was alarming because in late 1960, hospital acquired *Staphylococcal* strains were resistant to Penicillin (Thelma 1942). The decreased use of penicillin G against *Staphlococcus* and *Streptococcus* infections, opened new forum for production of semisynthetic penicillins (Kirby 1944; Kirby 1945; Medeiros 1984). In 1970, the penicillin with improved activity-ampicillin and amoxicillin, was introduced in market to treat infections caused by Enterobacteriaceae but ineffective against *Pseudomonas aeruginosa*.

The second generation of penicillin- the narrow spectrum methicillin was widely used to treat infections caused by *Staphylococcus epidermidis, Streptococcus pyogenes, Streptococcus pneumoniae* strains but soon within 20 years resistance to methicillin was conferred by *mecA* gene which encodes PBP-2a, located in *Staphylococcus aureus* chromosome (Matsuhashi et al., 1986). The narrow spectrum beta lactam antibiotic-oxacillin, nafcillin, cloxacillin, dicloxacillin were widely used against penicillin resistant *Staphylococcus aureus*. However, presently *methicillin-resistant Staphyloccus aureus* (MRSA) and *oxacillin-resistance Staph aureus* (ORSA) is global concern. Another narrow spectrum antibiotic, Temocillin (6-methoxy penicillin, Carboxypenicillin), β-lactamases resistant penicillin, analog of ticarcillin had greater stability than ticarcillin, active against infections caused by *Citrobacter* sp, *Escherichia coli, Klebsiella pneumoniae, Brucella abortus, Burkholderia cepacia, Salmonella* sp, *Pseudomonas aeruginosa* and *Proteus mirabilis* (Livmore et al., 2006). The extended-spectrum penicillin antibiotic, amdinocillin or mecillinam specifically binds to PBP2, active against infections caused by Gram negative bacteria (Nicolle 2000).

Table 1. List of penicillin derivatives in clinical practice

Sl No.	Antibiotic	Year of approval	Route of administration	Clinical Use
1	Penicillin G (Benzylpenicillin)	1946	Intramuscular or Intravenous	Pneumonia, Strep throat, Syphilis, Diptheria, Gas-gangere, Leptospirosis
2	Penicillin V (Phenoxymethyl penicillin)	1968	Oral	Strep throat, Otitis media, Cellulitis
3	Cloxacillin	1960	Oral, Intravenous	Pneumonia, septic arthritis
4	Methicillin	1960	Intravenous	Infections caused by *Staphylococcus*
5	Oxacillin	1962	Oral, Intravenous	Treatment of penicillin-resistant *Staphylococcus*
6	Ampicillin	1963	Oral, Intravenous	Group B *Streptococcal* infections, Respiratory tract infections, UTI, meningitis, endocarditis
7	Nafcillin	1970	Intravenous	*Staphylococcal* Infections
8	Amoxicillin	1972	Oral, intravenous	*Salmonella* infections, Chlamydia infections, UTI
9	Carbenicillin	1972	Oral	Active against Gram negative bacteria
10	Ticarcillin	1976	Intravenous	*Stenotrophomonas maltophiia*, Gram negative bacteria
11	Piperacillin	1981	Intravenous	Effective against Gram negative bacteria
12	Temocillin	1985	Intravenous	Treatment of multi drug resistant strains
13	Mecillinam	1978	Intravenous	Gram negative bacteria

The use of penicillin and its derivatives brought fruitful years to medicine, but the misuse of wonder drugs gave rise to antibiotic resistant strains. The strains of *S. pneumoniae* reported resistant to penicillin in 1967 (Hansman et al., 1974), beta lactamase-producing gonococci were isolated from England and United States in 1976 (Koornhof et al., 1992). In the period of ten years from 1976 to 1986, the gonococcal infections kept rising in Asian countries and a large outbreak of resistant gonococci affected

Durham city in North Carolina (Faruki et al., 1985). The resistance was due to mutations of these strains that modified the penicillin target PBP2 and expression of drug efflux system. This modification ruled out penicillin to treat gonococcal infections (Faruki et al., 1986; Hagman et al., 1995). By the year 2000, there was rise in hospital acquired beta lactam resistant *E. coli* strains. In 2008, New Delhi Metallo-beta-lactamase-1 (NDM-1) was first detected in *Klebsiella pneumoniae*, isolated from Swedish patient of Indian origin (Walsh et al., 2011). Further, NDM-1 was detected in India, Pakistan, United Kingdom, United States, Canada and Japan (Marilynn Marchione 2010; Madeleine White 2010; *Yuasa, Shino, 2010).*

The use of penicillin as monotherapy drug drastically decreased due to increasing beta lactamase producing organisms. More than 150 antibiotics have been introduced in market after the discovery of penicillin, but most of them are resistant to beta lactam group of drugs (Lobanovska and Pilla, 2017). The combination of β-lactamase inhibitors and antibiotics such as- ampicillin, amoxicillin, piperacillin is widely used against hospital acquired bacterial infections (Schaar et al., 2014). The rise in multi drug resistant strains associated with nosocomial infections-penicillin resistant *Enterococcus*, Vancomycin resistant *Enterococcus*, Multi-drug-resistant tuberculosis and carbapenem-resistant Enterobacteriaceae is alarming across the globe (Bartlett et al., 2013).

Cephalosporins

In 1950, a beta lactam antibiotic, Cephalosporin was derived from aerobic mold *Cephalosporium acremonium*, opened a new pathway for the development of novel antibiotics to treat infections caused by penicillinase-producing organisms especially *Staphylococcus aureus* (Newton and Abraham, 1956). The mode of action of cephalosporins is similar to beta-lactam antibiotic-penicillin, it inhibits cell wall synthesis resulting in the accumulation of nucleotides containing uridine-5-pyrophosphate and N-acetyl derivatives of muramic acid (Weinstein, 1980). The discovery of cephalosporin lead to three separate compounds: Cephalosporin P-mild activity against *Staphylococci, Corynebacterium* and *Clostridium*, Cephalosporin N-less activity against Gram positive bacteria and no clinical use, Cephalosporin C-wide range of activity against Gram-positive and Gram-negative bacteria. Further, resistance to compound Cephalosporin P emerged rapidly and was not used anymore in clinical practice (Abraham and Newton, 1961). The cephalosporins are clinically administered, either as parenteral or oral agents (Abraham, 1987). The cephalosporins administered parenteral was more effective than oral agents, it was used in some cases to replace oral penicillins in penicillin-allergic patients (Bush and Bradford 2016).

Cephalosporins are grouped into different generations on the basis of their antimicrobial properties, each generation is modified and improved to increase the spectrum of these antibiotics. The first generation cephalosporins were introduced in 1971 and was effective against Gram-positive bacteria including penicillinase producing *Staphylococci*. These drugs were used to treat mild to moderate skin infections caused by methicillin-susceptible *Staphlococcus aureus*, later these antibiotics were resistant to MRSA and *Enterococci* (Giordano et al., 2006; Sweetman, 2011). Table 2 represents the cephalosporins in current use. Cefazolin-first generation cephalosporin is still widely used for surgical porphylaxis, to treat abdominal infections (Sudo et al., 2014) and is effectively used as empiric therapy for upper respiratory tract infections in Japanese children (Abe et al., 2016). The second generation cephalosporins-cefaclor and cefuroxime are effective against organisms producing beta-lactamases mostly Gram-negative bacteria including Enterobacteriaceae. Cefuroxime, approved in 1983 was only member of cephalosporins II class with both oral and parenteral dosages but later the stability to β-lactamases was diminished by oral cephalosporins (Jacoby and Carreras 1990).

Table 2. List of available cephalosporin antibiotics

Sl. No.	Antibiotic	Year of Approval	Route of administration	Clinical use
1	Cephalexin (Cephalosporin-I)	1971	Oral	Middle ear Infections, bone, joint, skin, UTI, Pneumonia
2	Cefaclor (Cephalosporin-I)	1979	Oral	Septicaemia, Pneumonia, Meningitis
3	Cefazolin (Cephalosporin-I)	1973	Intravenous	Cellultis, Urinary tract infections, Pneumonia, Endocarditis
4	Cefuroxime (Cephalosporin-II)	1983	Oral, Intravenous	Pneumonia, Meningitis, sepsis, UTI, Lyme disease
5	Cefotaxime (Cephalosporin-III)	1981	Intravenous	Joint infections, Pelvic inflammatory disease, sepsis, gonorrhoea
6	Cefoperazone (Cephalosporin-III)	1982	Intravenous	Treating *Pseudomonas* bacterial infections
7	Ceftriaxone (Cephalosporin-III)	1984	Intravenous	Intra-abdominal infections, skin infections, UTI, gonoeehoea, pelvic inflammatory disease
8	Ceftazidime (Cephalosporin-III)	1985	Intravenous	Pneumonia, sepsis, UTI, *Pseudomonas* and *Vibrio* infection
9	Cefepime (Cephalosporin-IV)	1996	Intravenous	Nosocomial pneumonia, multi-drug resistant micro-organisms
10	Fosamil	2010	Intravenous	Methicillin resistant Staph aureus, Gram Positive bacteria
11	Ceftobiprole	2013	Intravenous	Hospital and community acquired pneumonia
12	Ceftolozane (Antipseudomonal cephalosporin)	2014	Intravenous	Urinary tract infections and intra-abdominal infections

The third generation cephalosporins possesses wide spectrum of activity when compared to other cephalosporins. These antibiotics are administered intramuscular or intravenously and continue to serve as lethal agents against Gram negative pathogens. Ceftriaxone was approved in 1982 and was effective against number of bacterial infections including *Serretia marcescens*, *Citrobacter* sp., *Nisseria meningitidis*, *Streptococcus pneumoniae*, beta-lactamase producing strains. The fourth generation cephalosporins are broad spectrum antibiotics with beta-lactamase stability and enhanced activity against Gram-positive and Gram-negative bacteria. Cefepime, an extended spectrum antibiotic was approved for medical use in 1994 (Fischer 2006). It was mostly used to treat moderate to severe nosocomial *pneumoniae*, multi-drug resistant microorganisms- *Pseudomonas aeruginosa* and bacterial strains causing infections in skin and urinary tract (Chapman and Perry, 2003). Cefepime tends to have lower MICs against-*Pseudomonas aeruginosa* >256 μg/mL, *Streptococcus pneumoniae* >8 μg/mL than other cephalosporins.

Carbapenems

Carbapenems are beta-lactam group of antibiotics and have a unique structure which provide protection against metallo-β-lactamases (MBL) and extended beta-lactamases (Knapp and English, 2001). These antibiotics are mostly used for the treatment of severe bacterial infections and multi-drug resistance strains. Carbapenems bind strongly to PBP2, PBP1a, PBP1b and PBP3 in Gram negative bacteria, eliminating the risk of resistance in carbapenems (Yang et al., 1995). In comparison to cephalosporins and penicillin, these antibiotics exhibit a broader spectrum of activity against Gram negative pathogens.

In 1970, thienamycin was naturally derived from *Streptomyces cattleya*- a potent broad-spectrum antibiotic effective against *Staphlococcal* infections, *Pseudomonas aeruginosa* and *Escherichia coli*. Due to chemical instability of thienamycin in aqueous solution, other derivatives were introduced for clinical practice. Imipenem is one such derivative, stability of imipenem was achieved by adding *N*-formimidoyl group to the 2-position. Imipenem has broad spectrum activity against Gram positive, Gram negative, aerobic, anaerobic bacteria, particularly among non-carbapenemase-producing enteric bacteria (Bradley et al., 1999; Kiratisin et al., 2012) but inactive against MRSA (Kesado et al., 1980). Later to prevent degradation of imipenem by human renal enzyme dehydropeptidase (DHP), it was combined with cilastatin-a DHP inhibitor (Kahan et al., 1983).

The derivatives of carbapenems including meropenems, ertapenems, doripenem, panipenem, biapenem have been introduced for better efficacy against bacterial infections. These derivatives of carbapenems are chemically more stable than imipenem allowing longer duration and prolonged infusion times (Cielecka-Piontek, et al., 2008; Prescott et al., 2011). The meropenem which is stable to mammalian DPH does not require cilastatin for profusion and effective against bacterial meningitis (Zhanel et al., 1998) due to its penetration into meninges (Dagan et al., 1994). The antibacterial activity of doripenem is similar to meropenem however, it is more effective against Gram-negative pathogens. The greater stability, prolonged infusions and less likely to produce seizures make doripenem an effective antibiotic than imipenem. In 2001, biapenem was approved in Japan and has similar efficacy like other carbapenems-meropenem and doripenem (Pei et al., 2014; Papp-Wallace et al., 2011). Tebipenem, a broad spectrum orally administered antibiotic is formulated as pivoxil ester for better absorption and bioavailability (Kato et al., 2010). The carbapenems used in clinical practice are listed in Table 3. Biapenem and tebipenem antibiotics are stable to hydrolysis of metallo- β-lactamases (MBL) compared to imipenem and meropenem (Neu et al., 1992, Inoue et al., 1995).

The development of resistance to carbapenems may be due to intrinsic or acquired resistance mechanisms, some pathogens attain resistance naturally among certain classes of antibiotics and few pathogens reduce the uptake of beta-lactam by altering their cell membrane porin channels (Codjoe and Donkor 2018). The resistance in clinically important pathogens is acquired by mutational events or by horizontal gene transfer (Meletis, 2016). Gram-negative bacteria such as *Pseudomonas aeruginosa*, develop resistance against carbapenems by preventing them to bind PBPs diminishing the permeability of their outer membrane (Bonomo and Szabo, 2006). The clinical pathogens develop enzyme- mediated resistance to carbapenems because these enzymes hydrolyze beta-lactam antibiotics called carbapenemases (Walsh, 2010). Among clinical pathogens, there are many plasmid mediated carbapenemases found in Enterobacteriaceae which can be easily spread among bacterial isolates. The number of Extended spectrum beta lactam (ESBL) genes have potential to transfer between organisms such as *Klebsiella, Enterobacter* sp, and *Escherischia coli*. The loss of OprD are the common mechanism of carbapenem resistance in *Pseudomonas aeruginosa, Enterobacter* sp. and *Klebsiella* against imipenem (Walsh 2011).

Table 3. List of carbapenems in clinical practice

Sl. No.	Antibiotic	Year of Approval	Route of Administration	Clinical Use
1	Imipenem	1985	Intravenous	Broad Spectrum Gram Positive and Gram-Negative bacteria
2	Meropenem	1996	Intravenous	Meningitis, Intra-abdominal infections, Pneumonia, sepsis
3	Ertapenem	2001	Intravenous	Pneumonia, UTI, Pelvic and abdominal Infections
4	Doripenem	2007	Intravenous	Hospital acquired *Pseudomonas aeruginosa*
5	Biapenem	2001	Intravenous	Active against anaerobic bacteria
6	Tebipenem	2009	Oral	Broad spectrum antibiotic

Monobactams

Monobactams also known as monocyclic beta-lactams, are family of antibiotic compounds produced by bacteria (Imada et al., 1981; Sykes et al., 1981). The variations in structure of monobactam led to the development of synthetic compound aztreonam. It was approved in 1986 in United States and effectively used to treat infections caused by Gram negative bacteria- *Citrobacter, Enterobacter, E. coli, Haemophilus, Klebsiella, Proteus* and *Serratia* sp. but ineffective against Gram positive and anaerobes (Mosby 2006). The mode of action is similar to penicillin, has high affinity towards PBP3 in comparison to PBP-1a resulting in cell lysis (Sykes et al., 1982). The combination of aztreonam and arbekacin or tobramycin has been suggested to increase the efficiency of this drug (Kobayashi et al., 1992). It was the only monobactam to gain regulatory approval for therapeutic use, however, the emergence of ESBLs and serine carbapenemases has limited the usage of aztreonam in clinical practice (Wang et al., 2014).

Antibiotic Resistance

The rapid emergence of resistant bacterial strains is worldwide major concern, the overuse and misuse of antibiotics have increased the risk of antibiotic resistance. The distribution of antibiotics prescribed to outpatients is uneven, in some cases antibiotics can last for months or years. In a recent survey, 30% of patients were prescribed one antibiotic per year in United Kingdom to treat bacterial infections (Shallcross et al., 2017). Sir Alexander Fleming warned regarding the overuse of antibiotics, "public will demand-then begin the era of abuses" (Spellberg and Gilbert 2014). The reason we are facing this problem globally is due to uneven and overuse of antibiotics which encourages the resistant strain to survive, exchange of resistance traits from one organism to another such as Gram-positive organisms to Gram-negative organisms and vice-versa (Levy 2000). The development of resistance in hospital acquired pathogens is an alarming problem especially in developing countries like India where the microbiological culture and sensitivity testing of each and every infection is not feasible (Thakuria and Lahon, 2013).

The extensive use of antibiotics in agriculture and livestock as growth supplements is efficiently contributing to antibiotic resistance. Antibiotics are used in livestock to improve overall health of animal and to produce high quality product (Micheal et al., 2014). In United States, 80% of antibiotics sold are used in livestock to improve growth and prevent infection in animals (Gross, 2013; Piddock, 2012). The antibiotics used in livestock reaches human intestine by consuming animal products, which kills or supress susceptible bacteria allowing resistant bacteria to survive causing serious various infections. A high number of antibiotic resistance strains have been reported in farm animals, farmers and individuals consuming meat. Another factor contributing to antimicrobial resistance, 80% of antibiotics used in animals are excreted in urine and stool, they are widely dispersed as fertilizer, groundwater affecting agricultural fields (Bartlett et al., 2013). Few antibiotics- such as tetracycline and streptomycin are used as pesticides and are sprayed on fruits in western countries, contributing to antimicrobial resistance (Golkar et al., 2014).

In past 20 years, the development of antibiotic resistance in multidrug resistance strains is a serious concern especially in hospital acquired pathogens including *Enterobacter* sp., *Escherichia coli, Haemophilus influenzae, Klebsiella pneumoniae, Pseudomonas aeruginosa, Proteus mirabilis, Staphylococcus aureus, Citrobacter* sp., *Serratia* sp., *Streptococcus pneumoniae* (Davies and Davies, 2010). The Centre of disease control (CDC) assessed the pattern of antibiotic resistance among pathogens and have categorized bacterial pathogens into:

1. Urgent threats

- Carbapenem resistance Enterobacteriaceae (CRE)
- Drug resistance *Nisseria gonorrhoeae*

2. Serious threats

- Vancomycin resistant *Enterococci*
- Extended spectrum beta-lactamase-producing *Enterobacteriaceae*
- Multi-drug resistance *Pseudomonas aeruginosa*
- Methicillin-resistant *Staphylococcus aureus*
- Multidrug resistant *Acinebacter*
- Drug resistant *Compylobacter*
- Multi-drug resistant tuberculosis

3. Concerning threats

- Vancomycin resistance *Staphlococcus aureus* (VRSA)

- Erythromycin-resistant Group A *Streptococcus*
- Clindamycin-resistant Group B *Streptococcus*

The increasing resistance to third and fourth generation of cephalosporins and carbapenems is worldwide concern. According to CDC, almost 78% skin and soft tissue infections in community is caused by MRSA and 64% of infections are hospital acquired infections. The beta-lactam antibiotics are found to be ineffective against methicillin-resistance Staph aureus (MRSA), also building resistance mechanism against ceftobiprole, fourth generation of cephalosporins (Banerjee et al., 2008; Rizvi et al., 2011). The only drug of choice against MRSA is vancomycin, linezolid or other antibiotics without a beta lactam ring. The Enterococci, common nosocomial pathogen accounts for 10% of the hospital acquired infections and also responsible for 16% of nosocomial urinary tract infections. It has been found that most of these strains emerging resistance to beta-lactams, aminoglycosides, however, cephalosporins which are currently in use are resistant to enterococci. Hospital acquired Enterococcus infection was treated with the combination of aminoglycoside and penicillin, but it has developed high level aminoglycoside resistance (HLAR) against vancomycin (Thakuria and Lohan 2013). CDC estimated that almost 30% hospital-acquired enterococcal infections per year are vancomycin resistant leading to 1.300 deaths each year.

The drug resistance *Streptococcus pneumoniae* can sometime cause life-threatening infections, claiming 7,000 deaths per year. It has developed resistance to penicillin class, erythromycin such as amoxicillin and azithromycin. The lack of new drugs against multi-drug resistant tuberculosis (TB) is making treatment of TB complex because of resistant drugs including isoniazid and rifampicin. The carbapenem resistance Enterobacteriaceae (CRE) have developed resistance against all available antibiotics which are reserved as "treatment of last resort" (Sengupta et al., 2013). The increasing resistance mechanisms in CRE is a major concern for future, the Klebsiella sp., reported from central line-associated bloodstream infections (CLABSIs) and catheter associated urinary tract infections (CAUTIs) were carbapenem resistant (Kalen and Guh, 2012).

The extended spectrum beta-lactamase (ESBL) confer an increased resistance towards most of the antibiotics available in market, the infections caused by ESBL producers developed resistance in some strains of *Klebsiella* and *E. coli* species in form of carbapenemases (Klebsiella producing carbapenemases and New Delhi metallo-β-lactamases). The Klebsiella producing carbapenemases (KPC) among Enterobacteriaceae, metallo-beta lactamases (MBL) were limited to few regions. However, the transmission between bacteria through mobile genes has spread the resistance worldwide and infections caused by these organisms is no more limited to hospitals (Hidron et al., 2008; Yong el al., 2009; Moellering, 2010). If we consider the present antibiotic resistance pattern, almost all beta-lactam antibiotics have become resistant to hospital associated microorganisms. The combination of beta-lactam and beta-lactamase inhibitor also cephalosporins are found to be ineffective against MRSA, AmpC and the MBL organisms. Beta-lactamase inhibitors such as Clavulanic acid, Sulbactam, Tazobactam, Avibactam and Relebactam are widely used to increase the potential of beta-lactam antibiotics. These inhibitors are naturally isolated from bacterial species and are combined with beta-lactam antibiotic including amoxicillin, ampicillin, piperacillin, imipenem and meropenem for efficient inhibition (Dong et al., 2014; Bush 2015; Moland et al., 2007).

Antibiotic-resistance infections is an economic burden to health care system and nation, also to patient and their families (Golkar et al., 2014). Most of antibiotic resistant infections are hospital acquired infections because of extensive use of antibiotics. When the first and second line of defensive antibiotics have become resistant, health care professionals may be forced to treat patient with strong or toxic antibiotic resulting in adverse effects (Lushnaik, 2014). The healthcare professionals must be aware of resistant flora which we are generating due to extensive and use of antibiotics. The organisms that are not inhibited by drug of choice such as beta-lactam drugs consequently overgrow by building resistant mechanisms. This might be the result of antibiotic usage and emergence of multi-drug resistant organisms (Dancer 2001).

CONCLUSION

The antibiotic resistance is a global concern, the overuse of antibiotics in clinical practice have increased resistance in Gram-negative pathogens. To overcome the present antibiotic resistance pattern, we need to create awareness among prescribers. It is very important to convince clinicians that they should prescribe particular antibiotics only according to patient's infection. There should be strict implementation of antibiotic policies in restricting beta-lactam supplies in hospitals, screening of antibiotic prescriptions by microbiologist before the administration of antibiotics in patients as well as livestock (Thakuria and Lahon, 2013). The lack of proper data collection, storage and microbiological sensitivity of resistant strains increases the complexity in beta-lactam resistance strains. Although, each year new antibiotics are introduced in market but there is an urgent need for the development of novel antibiotics because human race is fighting on third and fourth generation of cephalosporins. The efforts to implement new policies, development of novel antibiotics, creating awareness among health care professionals, upgrading laboratory services, regular check on nosocomial pathogens, the progressive steps in these areas might improve the antibiotic resistance pattern worldwide.

REFERENCES

Abe, Y., Wakabayashi, H., Ogawa, Y., Machida, A., Endo, M., Tamai, T., Sakurai, S., Hibino, S., Mikawa, T., Watanabe, Y (2016). Validation of cefazolin as initial antibiotic for first upper urinary tract infection in children. *Global Pediatric Health* 3: 1–7.

Abraham, E. P. (1987). *Cephalosporins* 1945-1986. *Drugs* 34:1-14.

Abraham, E. P. and Newton G. G. F. (1961). The structure of Cephalosporin C. *Biochemical Journal* 79:377.

Arias C. A and Murray, B. A (2009) Antibiotic resistance bugs in the 21st century-A clinical super challenge. *New England Journal of Medicine* 360:439-443.

Bartlett, J. G., Gilbert, D. N., Spellberg, B. (2013) Seven ways to preserve the miracle of antibiotics. *Clinical Infectious Diseases* 56(10):1445-50.

Banerjee, R., Gretes, M., Basuino, L., Strynadka, N., Chambers, H. F. (2008). In Vitro Selection and Characterization of Ceftobiprole-Resistant MethicillinResistant *Staphylococcus aureus*. *Antimicrobial Agents and Chemotherapy* 52(6):2089-96.

Bayles, K. W. (2000) The bactericidal action of penicillin: new clues to an unsolved mystery. *Trends Microbiology* 8:274-278.

Birnbaum, J., Kahan, F. M., Kropp, H., MacDonald, J. S. (1985). Carbapenems, a new class of beta-lactam antibiotics. Discovery and development of imipenem/cilastatin. *American Journal of Medicine*. 78 (6A): 3–21.

Bonomo, R and Szabo, D (2006) Mechanisms of multidrug resistance in *Acinetobacter* species and *Pseudomonas aeruginosa*. *Clinical Infectious Disease* 43: S49–S56.

Bradley, J. S., Garau, J., Lode, H., Rolston, K. V., Wilson, S. E., Quinn, J. P. (1999) *International Journal of Antimicrobial Agents* 2:93-100.

Bush, K. (2015) A resurgence of β-lactamase inhibitor combinations effective against multi-drug resistant Gram-negative pathogens. *International Journal of Antimicrobial Agents* 46:483-493.

Bush, K and Bradford, A. P. (2016) β-lactam and β-lactamase Inhibitors: An overview. Cold Spring Harbor Laboratory Press.

Centers for Disease Control and Prevention, Office of Infectious Disease Antibiotic resistance threats in the United States, 2013. Apr, 2013. Available at: http://www.cdc.gov/drugresistance/threat-report-2013. Accessed January 28, 2015.

Chapman, T. M. and Perry, C. M. (2003) Cefepime: a review of its use in the management of hospitalized patients with pneumonia. *American Journal of Respiratory and Critical Care Medicine* 2:75-107.

Cielecka-Piontek, J., Zajac, M., Jelinska, A (2008) A comparison of the stability of ertapenem and meropenem in pharmaceutical preparations in solid state. *Journal of Pharmaceutical and Biomedical Analysis* 46:52-57.

Codjoe, F. S and Donkor, E. S. (2017). Carbapenem Resistance: A Review. *Journal of Medical Sciences (Basel)* 6(1):1.

Dancer, S.J. (2001) The problem with cephalosporins. *Journal of Antimicrobials and Chemotherapy* 48:463-478

Dagan, R., Abramson, O., Leibovitz, E., Lang, R., Goshen, S., Greenberg, D., Yagupsky, P, Leiberman, A., Fliss, D. M. (1996) Impaired bacteriologic response to oral cephalosporins in acute otitis caused by pneumonococci with intermediate resistance to penicillin. *Pediatric Infectious Disease Journal* 15:980-985.

Davies, J and Davies, D (2010). Origin and evolution of antibiotic resistance. *Microbiology and Molecular Biology Reviews* 74(3):417-433.

Dong, X., Chen, F., Zhang, Y., Liu, H., Liu, Y., Ma, L. (2014) *In-vitro* activities of rifampin, colistin, sulbactam and tigecycline tested alone and in combination against extensively drug-resistant Acinetobacter-baumannii. *Journal of Antibiotics* 67:677-680.

Faruki, H., Kohmescher, R.N., Mc Kinney W.P., Sparling P.F. (1985). A community-based outbreak of infection with penicillin-resistant *Neisseria gonorrhoeae* not producing peniciliinase (chromosomally mediated resistance). *The New England Journal of Medicine* 313(10):607-611.

Faruki, H and Sparling, P.F. (1986) Genetics of resistance in a non-beta-lactamase-producing gonococcus with relatively high-level penicillin resistance. *Antimicrobial Agents Chemotherapy* 30(6):856-860.

Fleming, A. (1929). On the antibacterial action of cultures of a Penicillium, with special reference to their use in the isolation of *B. influenzae*.

Giordano, P. A, Elston, D., Akinlade, B. K., Weber, K., Notario, G. F., Busman, T. A., Cifaldi, M., Nilius, A. M. (2006). Cefdinir vs. cephalexin for mild to moderate uncomplicated skin and skin structure infections in adolescents and adults. *Current Medical Research and Opinion* 22: 2419–2428.

Goffin, C and Ghuysen, J. M. (1998). Multimodular Penicillin Binding Proteins: An enigmatic family of orthologs and paralogs. *Microbiology and Molecular Biology Reviews* 62:1079-1093.

Golkar, Z., Bagazra, O., Pace, D. G (2014) Bacteriophage therapy: a potential solution for the antibiotic resistance crisis. *The Journal of Infections in Developing Countries* 8(2):129–136. 13.

Gross, M (2013) Antibiotics in crisis. *Current Biology* 23(24):1063–1065.

Hagman, K., E., Pan, W., Spratt, B., G., Balthazar, J., T., Judd, R., C., Shafer, W., M. (1995) Resistance of Neisseria gonorrhoeae to antimicrobial hydrophobic agents is modulated by the mtrRCDE efflux system. *Microbiology* 141:611-22.

Hansman, D, Devitt, L., Miles, H., Riley, I. (1974). Pneumococci relatively insensitive to penicillin in Australia and New Guinea. *The Medical Journal of Australia*. 2(10):353-6.

Hare, R. (1982). New light on history of Penicillin. *Medical History* 26:1-24.

Hidron, A. I., Edwards, J. R., Patel, J., Horan, T. C., Sievert, D. M., Pollock, D. A. (2008) NHSN annual update: antimicrobial-resistant pathogens associated with healthcare-associated infections: annual summary of data reported to the National Healthcare Safety Network at the Centers for Disease Control and Prevention, 2006–2007. *Infection Control of Hospital and Epidemiology* 29:996-1011.

Imada, A., Kitano, K., Kintaka, K., Muroi, M and Asai, M (1981). Sulfazecin and isosulfazecin, novel β-lactam antibiotics of bacterial origin. *Nature* 289:590–591.

Jacoby, G. A. and Carreras, I (1990). Activities of beta-lactam antibiotics against Escherichia coli strains producing extended spectrum beta-lactamases. *Antimicrobial Agents and Chemotherapy* 34: 858–862.

Kahan, F. M., Kropp, H., Sundelof, J. G., Birnbaum, J. (1983). Thienamycin: Development of imipenem–cilastatin. *Journal of Antimicrobial Chemotherapy* 12: 1–35.

Kallen, A., Guh, A (2012) United States Centers for Disease Control and Prevention issue updated guidance for tackling carbapenemresistant enterobacteriaceae. *Euro Surveill*, 17(26).

Kato, K., Shirasaka, Y., Kuraoka, E., Kikuchi, A., Iguchi, M., Suzuki, H., Shibasaki, S., Kurosawa, T., Tamai, I. (2010). "Intestinal Absorption Mechanism of Tebipenem Pivoxil, a Novel Oral Carbapenem: Involvement of Human OATP Family in Apical Membrane Transport". *Molecular Pharmaceutics*. 7 (5): 1747–1756.

Kesado, T., Hashizume, T., Asahi, Y. (1980) Antibacterial activities of a new stabilized thienamycin N-formimidoyl thienamycin, in comparison with other antibiotics, *Antimicrobial Agents and Chemotherapy* 17:912-917.

Kiratisin, P., Chongthaleong, A., Tan, T., Y., Lagamayo, E., Roberts, S., Garcia, J., Davies, T. (2012) Comparative in-vitro activity of carbapenems against major Gram-negative pathogens: Results of Asia-Pacific surveillance from the COMPACT II study. *International Journal of Antimicrobial Agents* 39:311-319.

Kirby, W. M. (1944). Extraction of a highly potent inactivator from Penicillin resistant Staphylococci. *Science* 99:452-453.

Kirby, W. M. (1945) Bacteriostatic and lytic actions of Penicillin on sensitive and resistant Staphylococci. *Journal of Clinical Investigation* 24:165-169.

Knapp K. M., English B. K. (2001) Carbapenems. *Seminars in Pediatric Infectious Diseases* 12:175–185.

Kong, K. F., Schneper, L and Mathee, K (2010). Beta-lactam Antibiotics: From Antibiosis to Resistance and Bacteriology. *Acta Pathologica Microbiologica Immunologica Scandinavian* 118(1):1-36.

Koornhof, H. J., Wasas, A. and Klugman, K (1992) Antimicrobial resistance in *Streptococcus pneumoniae*: a South African perspective. *Clinical Infectious Diseases* 15(1):84-94.

Kobayashi, R., Murata, T., Yoshinaga, K. (1992) A follow-up study of 201 children with autism Kyushu and Yamaguchi areas, Japan. Journal of Autism and Development Disorders 22(3):395-411.

Levy, S. B (2000) The future of antibiotics: facing antibiotic resistance. *Clinical Microbiological Infections* 6:101-104.

Livermore, D. M., Hope, R., Fagan, E. J., Warner, N., Woodford, N. and Portz, N. (2006). Activity of temocillin vs prevalent ESBL and AMPC-producing Enterobacterriaceae from South-east England. *Journal of Antimicrobial Chemotherapy* 57(5):1012-4.

Lobanovska, M., and Pilla, G. (2017) Penicillin's discovery and antibiotic resistance: Lessons for the future? *Yale Journal of Biology and Medicine* 29:135-145.

Lushniak, B. D. (2014) Antibiotic resistance: a public health crisis. *Public Health Reports* 129(4):314–316.

Macfarlane, G. (1979) *Howard Florey, the making of a great scientist.* Oxford University Press, London.

Marchione, M (2010). "New drug-resistant superbugs found in 3 states." *Boston Globe.*

Mariya, L and Pilla, G (2017). Penicillin's discovery: Antibiotic resistance and Lessons for the future? *Yale Journal of Biology and Medicine* 90(1): 135-145.

Matsuhashi, M., Song, M. D., Ishino, F., Wachi, M., Doi, M., Inoue, M., (2000). Molecular cloning of the gene of a penicillin supposed to cause high resistance to beta lactam antibiotics in Staph aureus. *Antimicrobial agents and Chemotherapy* 44(6):1549-55.

Medeiros, A. A. (1984) β-lactamases. *British Medical Bulletin* 40:18-27.

Meletis, G. (2016). Carbapenem resistance: overview of the problem and future perspectives. *Therapeutic Advances in Infectious Diseases* 3(1):15–21.

Michael, C. A., Dominey-Howes, D., Labbate, M (2014) The antibiotic resistance crisis: causes, consequences, and management. *Front Public Health* 2:145.

Moellering, R. C. (2010) Jr. NDM-1—a cause for worldwide concern. *New England Journal of Medicine* 363: 2377-79.

Moland, E. S., Hong, S. G., Thomson, K. S., Larone, D. H., Hanson, N. D. (2007) A Klebsiella pneumoniae isolate producing at least eight different β-lactamases including an AmpC and KPC β-lactamase. *Antimicrobial Agents and Chemotherapy* 51:800-801.

Mosby's Drug Consult 2006 (16th ed.). Mosby, Inc.

Newton, G., G., F. and Abraham, E., P. (1956) Isolation of cephalosporin C, a penicillin like antibiotic containing D-α-aminoadipic acid. Biochemical Journal 62: 651-658.

Nicolle, L. E. (2000). "Pivmecillinam in the treatment of urinary tract infections." *Journal of Antimicrobial Chemotherapy 46: 35–39.*

Papp-Wallace, K. M., Endimiani, A., Taracila, M. A., Bonomo, R. A. (2011). "Carbapenems: past, present, and future." *Antimicrob. Agents Chemother.* 55 (11): 4943–60.

Pei, G., Yin, W., Zhang, Y., Wang, T., Mao, Y., Sun, Y (2014). "Efficacy and safety of biapenem in treatment of infectious disease: a meta-analysis of randomized controlled trials." *Journal of Chemotherapy.* 28 (1): 28–36.

Piddock, L. J (2012) The crisis of no new antibiotics—what is the way forward? *Lancet Infectious Diseases.* 2012;12(3):249–253.

Pottegard, A., Broe, A., Abenhus, R., Bjerrum, L., Hallas, J., Damkier, P. (2015). Use of antibiotics in children: A Danish nationwide drug utilization study. *The Pediatric Infectious Disease Journal* 34:16-22.

Prescott, WA Jr., Gentile, A., E., Nagel, J., L., Pettit, R., S. (2011) Continuous-infusion antipseudomonal β-lactam therapy in patients with cystic fibrosis. *Pharmacy and Therapeutics* 36: 723-763.

Rizvi, M. W., Shujatullah, F., Malik, A., Khan, H. M. (2011) Ceftobiprole - A novel cephalosporin to combat MRSA. *Eastern Journal of Medicine* 16:1-8.

Sauvage, E. F, Terrak, K. M., Ayala J. A. and Charlier P. (2008). The penicillin-binding-proteins: Structure and role in peptidoglycan synthesis. *FEMS Microbiology. Reviews* 32:234-258.

Schaar, V., Uddback, I., Nordstrom, T., Riesbeck, K. (2014). Group A *Streptococci* are protected from amoxicillin-mediated killing by vesicles containing β-lactamase derived from *Haemophilus influenzae. Journal of Antimicrobial Chemotherapy* 69:117-120.

Sengupta, S., Chattopadhyay, M. K., Grossart, H. P (2013) The multifaceted roles of antibiotics and antibiotic resistance in nature. *Frontiers in Microbiology*, 4:47.

Shallcross, L., Beckley, N., Rait, G., Hayward, A. and Peterson, I (2017) Antibiotic prescribing frequency amongst patients in primary care: a cohort study using electronic health record. *Journal Antimicrobial Chemotherapy* 72:1818-1824.

Shino, Y (2010). "Japan confirms first case of superbug gene". *The Boston Globe.*

Spellberg, B., Gilbert, D. N. (2014) The future of antibiotics and resistance: a tribute to a career of leadership by John Bartlett. *Clinical Infectious Diseases* 59 (suppl 2):S71–S75.

Sudo, T., Murakami, Y., Uemura, K., Hashimoto, Y., Kondo, N., Nakagawa, N., Ohge, H., Sueda, T. (2014). Perioperative antibiotics covering bile contamination prevent abdominal infectious complications after pancreatoduodenectomy in patients with preoperative biliary drainage. *World Journal of Surgery* 38: 2952–2959.

Sutherland, R. (1964). The nature of insensitivity of Gram negative bacteria towards Penicillins. *Journal of General Microbiology* 35:85-98.

Sweetman, S (2011). *The Complete Drug Reference.* London: Pharmaceutical Press.

Sykes, R. B., Bonner, D. P., Bush, K., Georgopapadakou, N. H., and Wells, J. S (1981). Monobactams–monocyclic β-lactam antibiotics produced by bacteria. *Journal of Antimicrobial Chemotherapy.* 8(Suppl. E):1–16.

Sykes, R., B., Boner, D., P., Bush, K., Georgopapadakou, N., H. (1982) Azthreonam (SQ 26, 776), a synthetic monobactam specifically active against aerobic Gram-negative bacteria. *Antimicrobial Agents and Chemotherapy* 21(1):85-92.

Tadataka, K., Terutaka, H., Yoshinari, A (1980). Antibacterial activities of a new stabilized thienamycin, N-formimidoyl thienamycin, in comparison with other antibiotics. "*Antimicrobial Agents and Chemotherapy.*" 17 (6): 912–7.

Thakuria, B. and Lahon, K (2013). The Beta lactam antibiotics as an empirical therapy in a developing country: An update on their current status and recommendations to counter the resistance against them. *Journal of Clinical and Diagnostic Research*, 7(6): 1207-1214.

Thelma, R (1942). Resistance of *Staphylococcus aureus* to the action of penicillin. *Experimental Biology and Medicine* 51:386-9.

Walsh, T. (2010) Emerging carbapenemases: a global perspective. *International Journal of Antimicrobial Agents* 36: S8–S14.

Walsh, T., Janis, W., David, L. M., Mark, A. T (2011) Dissemination of NDM-1 positive bacteria in the New Delhi environment and its implications for human health: an environmental point prevalence study. *The Lancet Infectious Diseases* 11(5): 355-362.

Wang, X., Zhang, F., Zhao, C., Wang, Z., Nichols, W. W., Testa, R., Li, H., Chen, H., He, W., Wang, Q. (2014). *In vitro* activities of ceftazidime–avibactam and aztreonam–avibactam against 372 Gram-negative bacilli collected in 2011 and 2012 from 11 teaching hospitals in China. *Antimicrobial Agents and Chemotherapy* 58: 1774–1778.

Watkins, R. R. and Bonomo A. R (2017). β-lactam antibiotics. Infectious Diseases 2:1203-1216.

Weinstein, J. A (1980). The Cephalosporins: Activity and clinical use. *Drugs* 19:137-154.

White, M (2010). "Superbug detected in GTA." *Toronto Star*.

Yang, Y, Bhachech, N., Bush, K (1995) Biochemical comparison of imipenem, meropenem and biapenem: Permeability, binding to penicillin-binding proteins, and stability to hydrolysis by b-lactamases. *Journal of Antimicrobial Chemotherapy* 35: 75–84.

Yong, D., Toleman, M. A., Giske, C. G., Cho, H. S., Sundman, K., Lee, K. (2009) Characterization of a new metallo- b-lactamase gene, bla NDM-1, and a novel erythromycin esterase gene carried on a unique genetic structure in Klebsiella pneumoniae sequence type 14 from India. *Antimicrobial Agents and Chemotherapy*, 53:5046-54.

Zhanel, G .G., Simor, A. E., Vercaigne, L., Mandell, L. (1998). "Imipenem and meropenem: Comparison of *in vitro* activity, pharmacokinetics, clinical trials and adverse effects." *Canadian Journal of Infectious Diseases*. 9 (4): 215–28.

INDEX

β

β-lactamase(s), vii, viii, ix, 2, 3, 4, 5, 6, 7, 8, 9, 10, 11, 12, 13, 14, 15, 16, 17, 18, 19, 20, 21, 22, 23, 24, 25, 26, 27, 28, 29, 30, 31, 33, 34, 36, 37, 39, 43, 44, 45, 46, 47, 48, 49, 50, 51, 52, 53, 54, 55, 56, 59, 60, 61, 62, 63, 64, 65, 66, 67, 69, 73, 84, 119, 120, 121, 123, 124, 125, 126, 130, 132, 134, 135, 137, 138, 139, 140, 143, 145, 147, 149, 150, 152, 153, 158, 160, 164, 165

A

A. baumannii (XDR-AB), viii, 120, 122, 130
A. baumannii complex (AB complex), viii, 119, 120, 121, 122, 123, 125, 131, 132, 134
Abraham, v, 143, 144, 149, 159, 165
acetic acid, 129
acetone, 144
acetylation, 124
acid, 31, 41, 53, 67, 125, 129, 145, 149, 158, 165

ADC, 124
adhesion, 126
adolescents, 161
adults, 161
adverse effects, 158, 167
Algeria, 25, 26, 30, 34, 78, 85, 140
allele, 38, 54, 57
AME, 124
aminoglycosides, 35, 121, 124, 125, 157
AmpC, v, vii, 1, 2, 3, 5, 6, 7, 8, 9, 10, 11, 12, 13, 14, 15, 16, 17, 18, 19, 20, 21, 22, 23, 24, 25, 26, 27, 28, 29, 30, 31, 32, 33, 35, 36, 37, 38, 40, 41, 42, 43, 44, 45, 46, 47, 48, 49, 50, 51, 52, 53, 54, 55, 56, 57, 58, 59, 60, 61, 62, 63, 64, 65, 66, 67, 68, 69, 70, 71, 72, 73, 74, 75, 78, 81, 83, 84, 86, 88, 89, 93, 94, 95, 96, 97, 100, 102, 103, 104, 105, 107, 108, 111, 112, 114, 115, 116, 125, 128, 129, 132, 137, 158, 164
antibiotic, vii, ix, 2, 36, 42, 68, 71, 74, 101, 102, 106, 111, 113, 122, 124, 126, 133, 137, 143, 146, 147, 148, 149, 152, 153, 154, 155, 156, 158, 159, 161, 162, 163, 164, 165

antibiotic resistance, vii, ix, 111, 113, 122, 124, 126, 133, 137, 143, 155, 156, 158, 159, 161, 162, 163, 164, 165
antimicrobial resistance, vii, viii, 3, 41, 54, 74, 75, 78, 84, 87, 96, 98, 101, 102, 109, 111, 113, 116, 117, 119, 122, 131, 132, 144, 156, 163
antimicrobial therapy, viii, 2, 75, 130, 134
Argentina, 70
Asia, 36, 39, 72, 97, 123, 124, 163
Asian countries, 35, 148
assessment, 81
autism, 163
autolysis, 146
avian, 72, 89, 95, 110
awareness, 159

B

bacteremia, 72, 107
bacteria, ix, 2, 3, 32, 33, 39, 42, 54, 56, 57, 58, 68, 69, 70, 71, 72, 73, 84, 87, 90, 97, 102, 106, 125, 129, 140, 143, 146, 147, 148, 150, 151, 152, 153, 154, 155, 158, 166, 167
bacterial infection, vii, ix, 2, 143, 144, 146, 149, 151, 152, 153, 155
bacterial pathogens, 156
bacterial strains, 152, 155
bacterium, 121, 123, 125
Bangladesh, 15, 21, 91
beef, 35, 37, 38, 39, 41, 58, 70, 95, 109
Belgium, 6, 7, 8, 9, 20, 28, 32, 33, 34, 35, 61, 64, 67, 79, 82, 135
beta lactamase, 108, 137, 146, 148, 149, 158
beta-lactams, ix, 100, 143, 144, 154, 157, 158, 163
bile, 166
bioavailability, 153
biosynthesis, ix, 143, 145
birds, 91

blood, 138
blood stream, 138
bloodstream, 91, 146, 157
bone, 151
Brazil, 18, 21, 25, 27, 28, 34, 35, 37, 70, 87, 95
breakdown, 129, 146
breeding, 72, 102
burn, 135, 138

C

carbapenem, ix, 123, 124, 129, 130, 132, 134, 137, 138, 139, 140, 141, 143, 144, 149, 154, 156, 157, 161, 163, 164
carbapenemases, 76, 120, 121, 123, 129, 130, 132, 134, 135, 140, 153, 155, 158, 167
cattle, vii, viii, 2, 3, 32, 35, 37, 39, 40, 41, 42, 43, 54, 56, 57, 75, 77, 83, 86, 87, 88, 89, 92, 94, 95, 100, 101, 102, 107, 109, 110, 111, 112, 113, 115, 116
CDC, 156, 157
cefazolin, 35, 40, 55, 159
cephalosporin, 23, 24, 33, 34, 37, 38, 40, 56, 76, 77, 78, 80, 84, 86, 88, 91, 94, 100, 106, 107, 114, 115, 144, 149, 150, 151, 159, 165
cephalosporin antibiotics, 151
chemical, 129, 152
chicken, 4, 14, 15, 18, 21, 30, 32, 33, 34, 35, 36, 37, 58, 70, 72, 74, 79, 83, 84, 85, 88, 91, 94, 96, 97, 98, 101, 102, 103, 104, 106, 108, 109, 111, 114, 116
children, 41, 150, 159, 163, 165
China, 4, 6, 7, 8, 9, 11, 12, 16, 18, 19, 21, 22, 26, 27, 32, 33, 35, 36, 44, 46, 57, 59, 60, 61, 63, 65, 66, 67, 77, 88, 89, 97, 98, 99, 112, 116, 118, 136, 141, 167
Chlamydia, 148
chromosome, 3, 125, 130, 147

classes, viii, 69, 120, 122, 123, 130, 153
classification, 121, 135, 145
clinical trials, 167
clone, 131, 136
cloning, 164
collaboration, 74
Colombia, 29
colonisation, 101
colonization, 73, 86, 90, 112, 136
color, iv
commercial, 93, 115
communities, 147
community, viii, 3, 68, 105, 111, 119, 120, 151, 157, 161
compilation, vii
complexity, 159
complications, 166
compounds, 73, 149, 154
Congress, iv
conjugation, 110, 125
consumption, 86, 90
contaminated water, 71
contamination, 42, 83, 166
control measures, ix, 73, 120, 126, 129, 134
controlled trials, 165
correlation, 132, 135
cost, 133
covalent bond, 146
covalent bonding, 146
covering, 166
cross-sectional study, 77
culture, 72, 144, 155
cystic fibrosis, 165
Czech Republic, 3, 11, 96, 136

D

damages, iv
data collection, 159
deaths, 157
DEFRA, 39, 84

degradation, 123, 153
Delta, 80
Denmark, 8, 10, 12, 14, 19, 20, 23, 24, 27, 37, 41, 44, 46, 48, 50, 51, 55, 58, 59, 60, 61, 62, 63, 64, 65, 70, 81, 84, 90, 91, 94, 116
derivatives, vii, ix, 143, 147, 148, 149, 152, 153
desiccation, 121
detection, viii, 2, 4, 33, 38, 40, 57, 69, 84, 89, 106, 124, 126, 127, 128, 129, 130, 134, 135
developing countries, ix, 143, 155
diarrhea, 37, 40, 56, 58, 115
diseases, 37, 57, 121, 139
disorder, 43
distribution, 4, 33, 77, 86, 155
diversity, 39, 79, 102, 106, 113, 133
drainage, 166
drug efflux, 149
drug resistance, 122, 126, 137, 152, 156, 157
drug treatment, 103
drugs, 145, 148, 149, 150, 157, 158

E

E. coli, viii, 2, 3, 5, 6, 7, 8, 9, 10, 11, 12, 13, 14, 15, 16, 17, 18, 19, 20, 21, 22, 23, 24, 25, 26, 27, 28, 29, 30, 31, 32, 33, 35, 36, 37, 38, 39, 41, 42, 43, 44, 45, 46, 47, 48, 49, 50, 51, 52, 53, 54, 55, 56, 57, 58, 59, 60, 61, 62, 63, 64, 65, 66, 67, 68,69, 70, 71, 72, 74, 75, 79, 80, 84, 91, 92, 95, 99, 114, 115, 149, 154, 158
E.coli, 23
East Asia, 113
EEA, 86
egg, 79
Egypt, 4, 23, 30, 37, 52, 76, 80, 86, 136

encoding, viii, 2, 3, 37, 38, 39, 40, 41, 42, 54, 55, 57, 58, 70, 77, 83, 91, 93, 95, 124, 131
England, 95, 106, 110, 136, 148, 164
enterobacteriaceae, vii, 2, 4, 5, 32, 36, 41, 42, 43, 55, 58, 59, 65, 68, 71, 76, 78, 81, 83, 87, 88, 96, 100, 101, 104, 105, 107, 108, 113, 118, 135, 147, 149, 150, 154, 156, 157, 158, 162
environment, vii, 2, 24, 30, 44, 45, 48, 49, 50, 52, 65, 70, 88, 93, 167
environmental contamination, 92
environments, 96, 101
enzyme, 33, 81, 107, 124, 131, 153
enzymes, 69, 95, 121, 124, 125, 128, 129, 132, 133, 145, 146, 153
epidemic, 83, 126
epidemiology, 3, 74, 83, 85, 91, 94, 123, 133, 134, 136, 139
ESCs, 78
ester, 153
Estonia, 58, 66, 80
Europe, 33, 36, 69, 81, 123, 135
European Union, 38, 56, 69, 87
evidence, 69, 71, 110, 115
evolution, 82, 133, 161
exposure, 71, 72, 85
extended-spectrum β-lactamases (ESBL), v, vii, 1, 2, 3, 5, 6, 7, 8, 9, 10, 12, 13, 18, 25, 27, 30, 32, 33, 34, 35, 36, 37, 38, 39, 40, 41, 42, 43, 44, 46, 47, 48, 51, 53, 54, 55, 56, 57, 58, 59, 60, 61, 65, 66, 68, 69, 70, 71, 73, 74, 75, 78, 79, 81, 83, 84, 85, 87, 88, 90, 91, 92, 93, 94, 95, 96, 97, 98, 99, 102, 104, 106, 107, 108, 109, 110, 111, 113, 114, 118, 129, 132, 154, 158, 163
extracts, 144

F

families, 54, 81, 90, 158
farm environment, 65, 112
farmers, 68, 72, 75, 84, 85, 87, 90, 155
farms, viii, 2, 17, 19, 38, 39, 40, 42, 55, 56, 61, 68, 69, 72, 75, 76, 77, 79, 83, 84, 85, 87, 88, 89, 90, 93, 94, 96, 97, 101, 102, 103, 104, 106, 109, 110, 112, 114
FDA, 37, 87
feces, 32, 33, 34, 38, 39, 55, 56, 116
fluoroquinolones, 35, 71, 115, 121, 131
food, vii, 2, 3, 9, 32, 33, 35, 36, 40, 41, 46, 57, 69, 70, 73, 74, 76, 77, 78, 81, 84, 85, 87, 88, 89, 92, 94, 97, 99, 100, 101, 102, 105, 107, 112, 113, 114, 115, 116, 117, 118
Food and Drug Administration, 1, 75
food-producing animals, vii, viii, 2, 3, 33, 35, 36, 41, 57, 69, 70, 73, 77, 78, 92, 101, 102, 113, 114
Food-Producing Animals, v, 71, 104
formation, 121, 125, 139
France, 5, 6, 7, 10, 33, 34, 39, 44, 45, 47, 48, 49, 55, 56, 57, 59, 69, 70, 79, 82, 83, 89, 91, 95, 100, 101, 115, 136, 138
fruits, 156
fungi, ix, 143

G

Gabon, 22, 108
gene combinations, 57
gene transfer, 77, 153
genes, vii, viii, 2, 3, 4, 32, 33, 34, 35, 36, 39, 41, 54, 55, 56, 57, 68, 69, 71, 72, 73, 74, 75, 77, 78, 87, 88, 94, 95, 97, 98, 99, 101, 104, 108, 109, 113, 114, 117, 120, 121, 123, 124, 126, 130, 131, 132, 133, 134, 135, 138, 140, 154, 158
genetic diversity, 138

genome, 75, 126, 133, 139
genomics, 130
genotype, 35, 134
genotyping, 129, 134, 140
Germany, 3, 15, 18, 22, 25, 29, 30, 36, 37, 42, 46, 47, 50, 52, 53, 54, 55, 57, 58, 62, 64, 65, 66, 67, 69, 70, 83, 86, 96, 107, 108, 115
Great Britain, 77, 106
growth, 144, 155
Guangdong, 99
Guangzhou, 141
guidance, 162
guidelines, 128
Guinea, 162

H

health, viii, 3, 73, 105, 119, 120, 121, 155, 158, 159, 166
health care, viii, 119, 120, 121, 158, 159
health care professionals, 158, 159
health care system, 158
history, 162
Hong Kong, 35, 85, 92
horses, 69, 94
human, vii, 2, 35, 39, 55, 68, 71, 72, 73, 74, 75, 79, 89, 91, 103, 110, 112, 115, 121, 138, 153, 155, 159, 167
human exposure, 73
human health, 73, 74, 75, 167
Hunter, 46, 93
hydrolysis, 146, 153, 167
hydroxyl, 124
hygiene, 75

I

identification, 126, 133
in vitro, 167
incidence, 38, 125

indentation, 128
India, 3, 24, 26, 29, 30, 32, 36, 42, 51, 52, 54, 65, 72, 81, 94, 96, 100, 108, 119, 122, 132, 137, 139, 140, 149, 155, 167
individuals, 155
Indonesia, 51, 111
industries, 75
industry, 68, 74
infection, ix, 55, 73, 78, 87, 91, 120, 126, 129, 133, 134, 136, 146, 151, 155, 157, 159, 161
inflammatory disease, 151
inhibition, 127, 128, 158
inhibitor, 71, 115, 129, 145, 153, 158, 160
insertion, viii, 2, 39
intensive care unit, 90
intervention, ix, 120, 130, 134
Iran, 108, 139
Ireland, 6, 10, 36, 80
isolation, 55, 70, 74, 114, 115, 161
Israel, 51, 76
Italy, 8, 10, 12, 15, 23, 24, 34, 36, 42, 45, 46, 54, 55, 87, 88, 89

J

Japan, 4, 6, 7, 12, 15, 16, 17, 22, 30, 32, 33, 34, 35, 38, 39, 40, 43, 46, 47, 48, 49, 50, 51, 55, 56, 58, 61, 62, 63, 70, 77, 91, 92, 94, 101, 104, 108, 109, 110, 111, 140, 149, 153, 163, 166
Java, 6, 11, 13

K

Korea, 12, 13, 14, 23, 24, 25, 31, 33, 35, 36, 50, 51, 57, 60, 63, 64, 66, 67, 68, 90, 95, 99, 112
Kuwait, 135

L

labeling, 102
lead, vii, viii, 2, 149
leadership, 166
light, 162
liver, 40, 57
livestock, 33, 37, 57, 91, 107, 155, 159
longitudinal study, 85, 114
lymph, 40
lymph node, 40
lysis, 144, 154

M

majority, 120, 129
Malaysia, 26, 77
mammals, 96
management, 71, 126, 160, 164
manure, 20, 88
mastitis, 42, 54, 55, 58, 83, 86, 88, 99, 104, 108
matter, iv, 75
meat, vii, 2, 4, 5, 6, 7, 8, 9, 11, 12, 13, 14, 18, 19, 23, 24, 25, 26, 27, 28, 29, 30, 31, 33, 34, 36, 37, 38, 40, 42, 43, 45, 46, 58, 59, 60, 68, 69, 70, 72, 73, 76, 77, 78, 79, 81, 83, 85, 86, 88, 89, 91, 92, 94, 95, 96, 97, 101, 104, 108, 109, 111, 113, 114, 116, 118, 155
media, 41, 148
median, 13, 47, 62
medical, 120, 121, 152
medicine, 74, 144, 148
meninges, 153
meningitis, 146, 148, 153
meta-analysis, 165
metallo β-lactamase (MBL), viii, 120, 122, 123, 125, 126, 127, 128, 129, 131, 132, 135, 137, 138, 140, 152, 153, 158
Mexico, 7, 40, 45, 60, 102, 117
mice, 144
microorganisms, 73, 128, 129, 135, 152, 158
middle ear infection, 146
Middle East, 136
Min, Ho Chi, 103
misuse, 148, 155
mold, 144, 149
molecular structure, ix, 135, 143
Moon, 95, 98
mutation, 31, 41, 53, 67, 124, 133
mutations, 38, 39, 124, 133, 149

N

National Survey, 76
Netherlands, 3, 5, 6, 11, 13, 14, 17, 19, 20, 21, 23, 24, 34, 36, 37, 52, 53, 67, 68, 69, 70, 72, 74, 84, 91, 101, 104, 108, 114
New Delhi Metallo β-lactamase (NDM), viii, 66, 120, 121, 125, 128, 130, 132, 135, 136, 137, 139, 149, 164, 167
New England, 160, 161, 164
Nigeria, 4, 12, 28, 37, 100, 104
Nile, 80
North America, 34, 38
nosocomial pneumonia, 152
nucleotides, 149

P

Pacific, 163
paediatric patients, 147
Pakistan, 137, 149
pathogens, 93, 102, 122, 123, 130, 133, 134, 139, 151, 152, 153, 155, 156, 159, 160, 162, 163
pathway, 149
pathways, 126
PCM, viii, 120, 121, 125, 130, 132
PCR, viii, 39, 120

pelvic inflammatory disease, 151
penicillin, 122, 144, 146, 147, 148, 149, 152, 154, 157, 160, 161, 162, 164, 165, 167
peptide, 145
peptide chain, 145
permeability, 129, 146, 153
permission, iv
pharmaceutical, 160
pharmacokinetics, 167
phenotype, 122
phenotypes, 139
phosphorylation, 124
pigs, vii, viii, 2, 3, 32, 35, 41, 55, 57, 58, 61, 68, 71, 75, 76, 77, 84, 85, 86, 89, 90, 92, 93, 94, 97, 103, 104, 106, 108, 110, 113, 116
plants, 39, 118
plasmid, 38, 41, 54, 55, 57, 58, 68, 72, 75, 78, 81, 82, 83, 87, 88, 89, 91, 98, 100, 105, 109, 114, 115, 116, 117, 154
plasticity, 133
pneumonia, 151, 160
point mutation, 133
Poland, 16, 32, 58, 70, 114, 135
policy, 74
population, 72, 87, 107, 145
Portugal, 3, 8, 9, 10, 32, 36, 57, 60, 61, 63, 68, 69, 81, 100, 106
poultry, vii, viii, 2, 3, 5, 32, 33, 34, 68, 69, 70, 71, 72, 74, 75, 76, 77, 78, 79, 80, 82, 84, 85, 86, 87, 89, 91, 93, 94, 96, 97, 98, 100, 102, 103, 108, 109, 110, 111, 114, 115, 116, 117, 118
preparation, iv
prevention, 133
producers, viii, 2, 4, 18, 32, 35, 36, 39, 40, 42, 54, 55, 56, 57, 68, 69, 70, 72, 74, 75, 81, 109, 120, 124, 125, 128, 130, 134, 158
professionals, 158
prophylaxis, 71

protection, 152
proteins, 122, 124, 126, 129, 165, 167
Pseudomonas aeruginosa, 29, 32, 135, 137, 138, 144, 147, 152, 153, 154, 156, 160
public health, 3, 73, 75, 164
pumps, 122, 124, 126
pyrophosphate, 149

R

race, 159
recommendations, iv, 73, 166
recovery, 3, 39, 102
rehabilitation, 136
researchers, 39, 42
residue, 145
residues, 42, 106, 146
response, 161
restriction fragment length polymorphis, 39
retail, 13, 35, 38, 40, 71, 76, 77, 83, 85, 92, 94, 95, 96, 97, 102, 104, 109, 111, 113, 114, 116, 117, 118
rights, iv
risk, vii, viii, 2, 3, 70, 71, 73, 78, 89, 93, 101, 107, 108, 112, 115, 152, 155
risk factors, vii, viii, 2, 71, 72, 73, 78, 89, 93, 101, 107, 108
room temperature, 144
routes, 73, 133

S

safety, 73, 165
salmonella, 87, 115, 117
Salmonella, vii, 2, 3, 5, 6, 7, 10, 13, 24, 28, 30, 32, 34, 37, 40, 41, 56, 68, 69, 75, 77, 79, 80, 81, 82, 84, 85, 86, 87, 88, 90, 91, 92, 93, 94, 95, 97, 100, 101, 105, 106, 107, 109, 110, 111, 112, 113, 114, 115, 116, 117, 118, 144, 147, 148
Saudi Arabia, 140

second generation, 147, 150
sensitivity, 129, 155, 159
sepsis, 146, 151, 154
septic arthritis, 148
sequencing, 126, 133, 135
serine, vii, viii, 120, 123, 155
services, iv, 159
sewage, 101
showing, viii, 69, 120, 121, 122, 127, 128
Singapore, 27, 98
skin, 150, 151, 152, 157, 161
solid state, 161
solution, 152, 162
South Africa, 163
South America, 34, 70, 102, 123, 124
South Korea, 82, 95, 97, 98, 106, 109, 116
Spain, 3, 5, 6, 7, 10, 11, 15, 19, 26, 32, 33, 34, 38, 39, 55, 56, 57, 59, 61, 68, 69, 72, 80, 83, 85, 86, 103, 107, 113
species, ix, 3, 13, 68, 69, 89, 106, 115, 120, 121, 122, 132, 134, 143, 158, 160
sperm, 102
spleen, 40
Spring, 160
stability, 146, 147, 150, 152, 153, 160, 167
stakeholders, 75
states, 164
storage, 159
structure, 145, 152, 154, 159, 161, 167
subgroups, 146
substitution, 81, 131
Sun, 1, 98, 99, 113, 165
supplementation, 93
supplier, 101
surveillance, ix, 75, 86, 111, 120, 126, 129, 134, 163
susceptibility, 84, 88, 97, 112, 129, 132, 136
Sweden, 14, 17, 21, 50, 64, 70, 73, 79, 86, 111
Switzerland, 4, 15, 22, 28, 30, 33, 34, 36, 37, 38, 46, 48, 49, 51, 56, 61, 62, 67, 86, 88, 91, 96

syndrome, 58
synthesis, 145, 149, 165

T

Taiwan, 5, 59, 66, 113, 116
target, 149
teams, 133
techniques, 135
TEM, vii, 2, 5, 6, 7, 8, 9, 10, 11, 12, 13, 14, 15, 16, 17, 18, 19, 20, 21, 22, 23, 24, 25, 26, 27, 28, 29, 30, 31, 33, 34, 36, 37, 39, 40, 41, 43, 44, 45, 46, 47, 48, 49, 50, 51, 52, 53, 54, 56, 57, 59, 60, 61, 62, 63, 64, 65, 66, 67, 68, 69, 72, 81, 92, 106, 107, 109, 124, 125
testing, 140, 155
Thailand, 64
therapeutic use, 155
therapy, ix, 120, 126, 130, 134, 150, 162, 165, 166
third-generation cephalosporin, 3, 86, 88, 114
threats, 156, 160
time frame, 133
tissue, 157
tonsils, 113
toxin, 70, 77, 95, 108, 115
trade, 108
traits, 107, 155
transmission, 3, 68, 72, 73, 74, 75, 118, 121, 158
treatment, vii, 2, 68, 74, 126, 134, 152, 157, 165
tuberculosis, 149, 156, 157
Turkey, 42, 53, 54

U

United Kingdom, 149, 155

United States, 90, 112, 116, 117, 148, 154, 155, 160, 162
upper respiratory tract, 150
urinary tract, 37, 39, 56, 57, 113, 139, 146, 152, 157, 159, 165
urinary tract infection, 37, 39, 56, 57, 113, 139, 146, 157, 159, 165
urine, 125, 156
US Department of Health and Human Services, 75
USA, 8, 11, 16, 31, 35, 36, 37, 38, 40, 41, 43, 44, 45, 46, 53, 59, 63, 69, 72, 85, 104, 123

W

Wales, 95, 106, 110, 111
war, 136
waste, 42, 74, 106
water, 20, 21, 71, 74, 88
welfare, 74
wholesale, 35
withdrawal, 71
workers, 74, 103, 112, 128, 144
World War I, 144
worldwide, 3, 120, 155, 157, 158, 159, 164

Related Nova Publications

PLANT GROWTH PROMOTING MICROORGANISMS: MICROBIAL RESOURCES FOR ENHANCED AGRICULTURAL PRODUCTIVITY

EDITORS: Niranjan S. Raj and A. C. Udayashankar

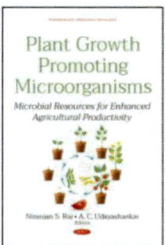

SERIES: Microbiology Research Advances

BOOK DESCRIPTION: Plant growth promoting microorganisms (PGPM) have gained acceptance and importance due to their dual benefits of promoting plant growth in addition to managing plant pests and diseases and are extensively used as microbial inoculants in improving agricultural productivity.

HARDCOVER ISBN: 978-1-53615-776-5
RETAIL PRICE: $230

INFECTIOUS DISEASES: RESPONSE, RECOVERY AND TRENDS

EDITOR: Philippe Georges

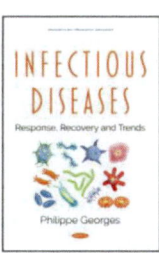

SERIES: Microbiology Research Advances

BOOK DESCRIPTION: An infectious disease threat is unique because of the transmissibility of diseases and the mobility of human populations. Infectious disease threats in recent years—such as Zika and Ebola outbreaks—have heightened the United States' attention to future potential threats, and raised questions about the nation's preparedness and response capabilities.

HARDCOVER ISBN: 978-1-53616-401-5
RETAIL PRICE: $230

To see a complete list of Nova publications, please visit our website at www.novapublishers.com

Related Nova Publications

ENTER THE WORLD OF MICROBIOLOGY: INTERVIEWS ABOUT THE WORLD'S MOST FAMOUS MICROBIOLOGISTS

AUTHORS: Manuel Varela and Michael F. Shaughnessy

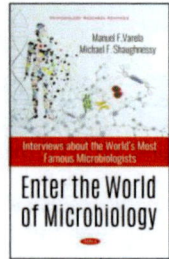

SERIES: Microbiology Research Advances

BOOK DESCRIPTION: Enter the world of microbiology: A dimension not only of viruses and bacteria, but also of contributions to medicine, health, and well-being. Enter the world of the most famous discoveries of scientists in the realm of biology from all over the world.

HARDCOVER ISBN: 978-1-53615-168-8
RETAIL PRICE: $195

To see a complete list of Nova publications, please visit our website at www.novapublishers.com